My Book

This book belongs to

Cover Design by :
Gowri Vemuri

First Edition :
May, 2019

Second Edition:

April, 2021

Author :
Gowri Vemuri

Edited by :
Raksha Pothapragada
Ritvik Pothapragada
Questions: mathknots.help@gmail.com

NOTE : VDOE is neither affiliated nor sponsors or endorses this product.

Dedication

This book is dedicated to:

My Mom, who is my best critic, guide and supporter.

To what I am today, and what I am going to become tomorrow,

is all because of your blessings, unconditional affection and support.

This book is dedicated to the

strongest women of my life ,

my dearest mom

and

to all those moms in this universe.

G.V.

Visit www.a4ace.com

Also available more time based practice tests on subscription

The Virginia Board of Education and Virginia Department of Education (VDOE) have developed the Virginia Assessment Program (VAP) to measure and evaluate students' academic progress in the Standards of Learning (SOLs). The SOLs indicate Virginia's expectations for what students should know and be able to do in the subject areas of reading, writing, mathematics, science, and history/social science.

Students in grades 3-12 to take the Standards of Learning (SOL) assessments each year. Some of the tests are required by all students each year, and others are required only at specific grade levels. Additionally, with the removal of some SOL tests in recent years, the VDOE assigned the responsibility of the creation and administration of alternate, performance - based assessments on local divisions. Student scores from these tests determine a school's and the division's state accreditation and measures progress toward meeting federal targets.

Virginia Standards of Learning (SOL) tests are generally given online unless a student has an identified and documented need to be assessed using paper,pencil format. The test question format is typically multiple choice, and each test contains some technology enhanced items.

GRADE 3-8 STANDARDS OF LEARNING(SOL) TESTS

GRADE 3	GRADE 4	GRADE 5	GRADE 6	GRADE 7	GRADE 8
	VIRGINIA STUDIES				WRITING
MATH	MATH	MATH	MATH	MATH	MATH
READING	READING	READING	READING	READING	READING
		SCIENCE			SCIENCE

NOTE : VDOE is neither affiliated nor sponsors or endorses this product.

END OF COURSE STANDARDS OF LEARNING (SOL) TESTS

GRADE 9	GRADE 10	GRADE 11
ALGEBRA I	GEOMETRY	ALGEBRA II
EARTH SCIENCE	BIOLOGY	CHEMISTRY
WORLD HISTORY I	WORLD HISTORY II	VIRGINIA & US HSTORY
		WORLD GEOGRAPHY
		ENGLISH : READING
		ENGLISH : WRITING

Any Student taking one of the courses listed here is expected to take the corresponding end-of-course SOL test. The grade levels depicted here represent grade level at which students typically participate in these courses.

NOTE : VDOE is neither affiliated nor sponsors or endorses this product.

SOL Test Scoring and Performance Reports:

Standards of Learning assessments in English reading, mathematics, science and history/social science are made up of 35-50 items or questions that measure content knowledge, scientific and mathematical processes, reasoning and critical thinking skills. English writing skills are measured with a two-part assessment that includes multiple-choice items and an essay.

Student performance is graded on a scale of 0-600 with 400 representing the minimum level of acceptable proficiency and 500 representing advanced proficiency. On English reading and mathematics tests, the Board of Education has defined three levels of student achievement: basic, proficient, and advanced, with basic describing progress towards proficiency.

Performance Achievement Levels:

- The achievement levels for grades 3-8 reading and mathematics tests are: *Pass/Advanced, Pass/Proficient, Fail/Basic,* and *Fail/Below Basic.*

- The achievement levels for science tests, history tests, and End-of-Course (EOC) tests* are: *Pass/Advanced, Pass/Proficient*, and *Fail/Does Not Meet.*

- The EOC Writing (2010 SOL) test, EOC Reading (2010 SOL) test, and EOC Algebra II (2009 SOL) test have an achievement level of *Advanced/College Path* in place of the *Pass/Advanced* achievement level.

NOTE : VDOE is neither affiliated nor sponsors or endorses this product.

INDEX

FORMULA SHEET

1. Area of a triangle

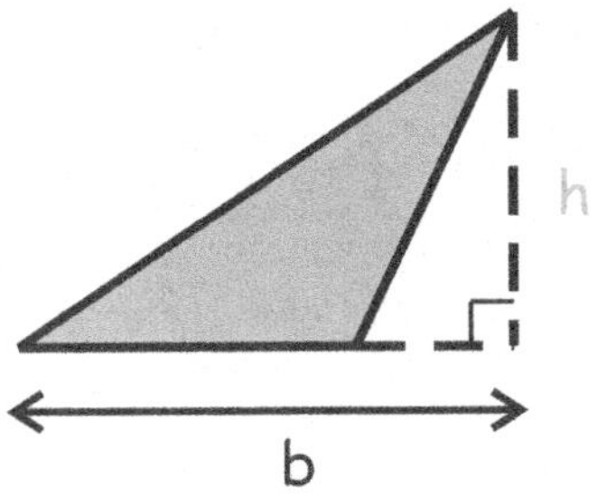

$A = \frac{1}{2} bh$

2. Area of a parellelogram

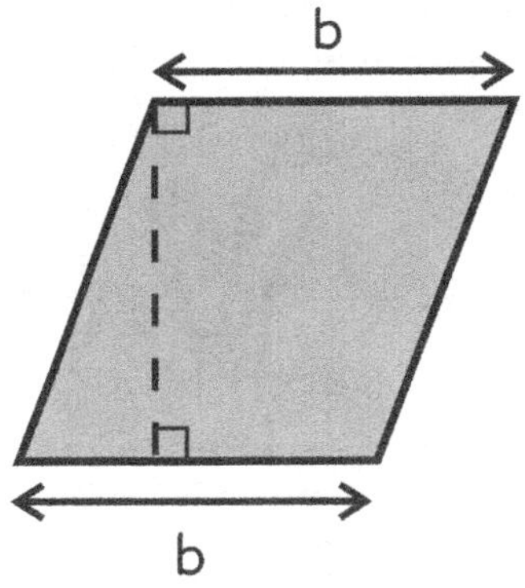

$A = bh$

3. Volume and Surface area of a cuboid

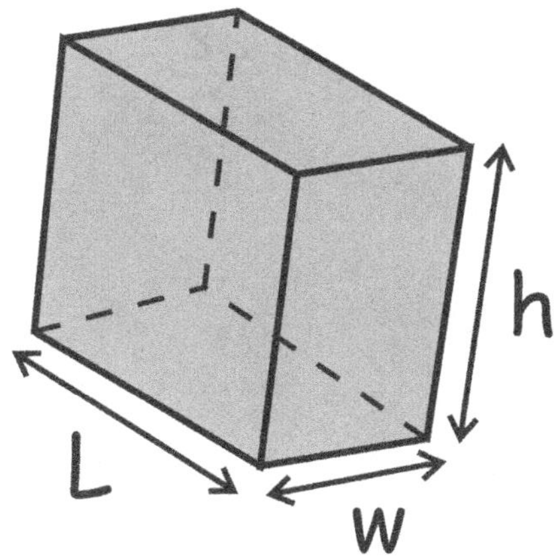

$V = lwh$

$S.A = 2(lw + lh + wh)$

4. Perimeter and Area of a Square

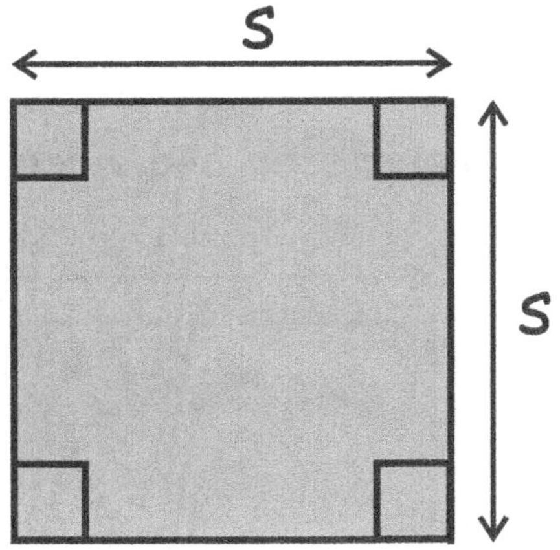

$P = 4s$

$A = s^2$

5. Area of a Trapezium

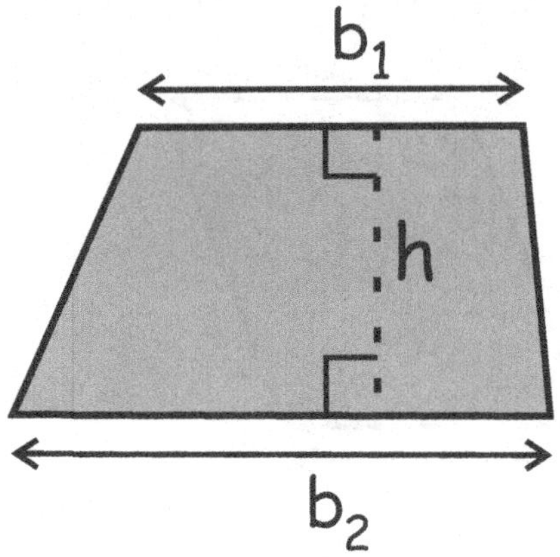

$$A = \frac{1}{2} h(b_1 + b_2)$$

6. Circumference and Area of a Circle

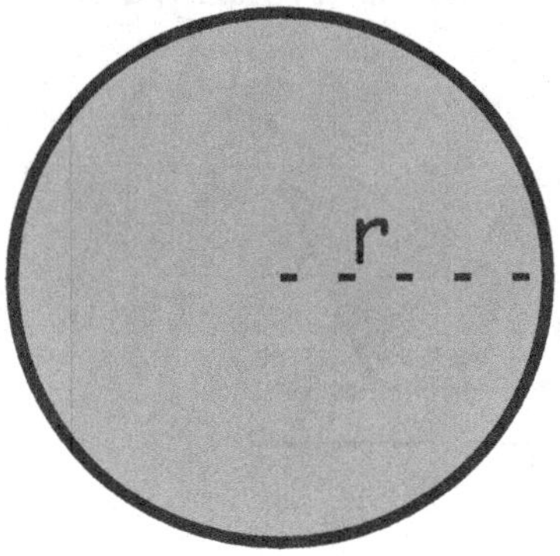

$$c = 2\Pi r$$
$$A = \Pi r^2$$

pi

$$\Pi = 3.14$$
$$\Pi = \frac{22}{7}$$

7. Perimeter and Area of a Rectangle

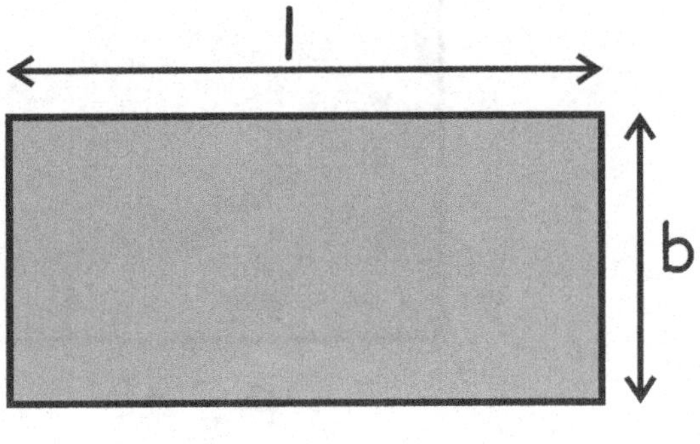

Area = l b x

Perimeter = 2(l + b)

8. Acute angle

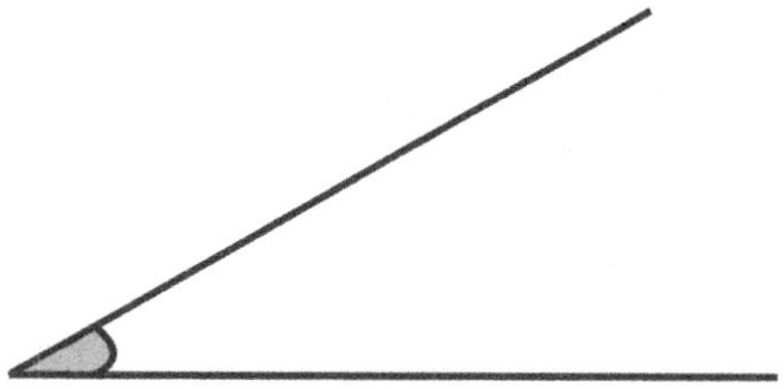

Any angle less than 90^0
is called as acute angled

9. Right angle

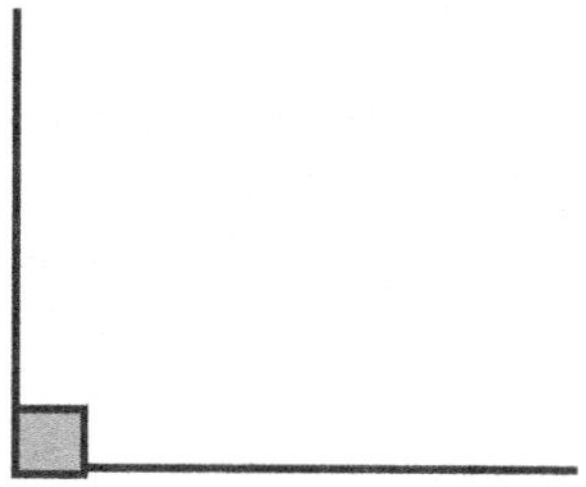

An angle equal to 90^0
is called as right angled

10. Obtuse angle

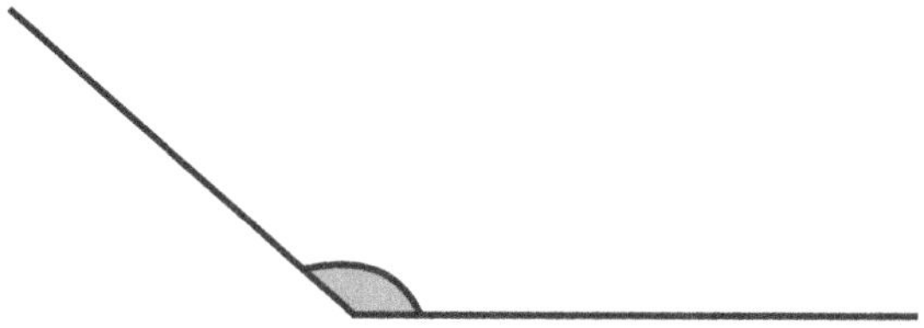

Any angle greater than 90^0
is called as obtuse angled

11. Perimeter of a polygon KLMNOP

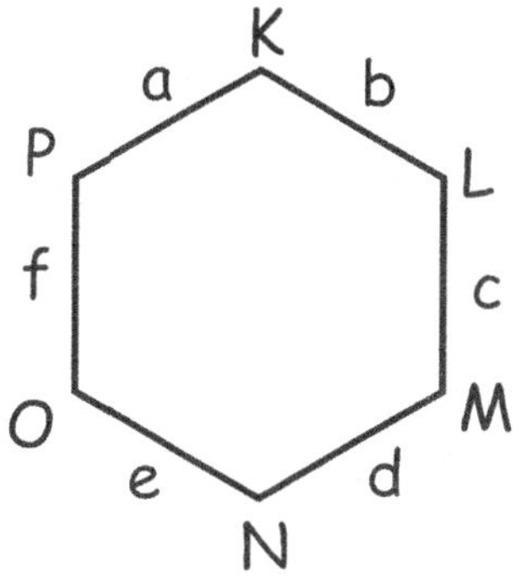

Perimeter =
(a + b + c + d + e + f) units

Abbreviations

milligram	mg
gram	g
kilogram	kg
milliliter	mL
liter	L
kiloliter	kL
millimeter	mm
centimeter	cm
meter	m
kilometer	km
square centimeter	cm^2
cubic centimeter	cm^3

year	yr
month	mon
hour	hr
minute	min
second	sec

volume	V
total Square Area	S.A
area of base	B
ounce	oz
pound	lb
quart	qt
gallon	gal.
inches	in.
foot	ft
yard	yd
mile	mi.
square inch	sq in.
square foot	sq ft
cubic inch	cu in.
cubic foot	cu ft

Grade 3
SOL
Practice Test - 1

1. Fantasy Music Club sold 6,439 tickets for their music concert. What is 6,439 rounded to the nearest hundred ?

   (A) 6,600 (B) 6,470 (C) 5,480 (D) 6,400

2. Which of the below shows "Nine hundred thousand, nine hundred nine" in standard form ?

   (A) 900,909 (B) 900,919

   (C) 900,400 (D) 9,000,909

3. Eighteen lolly pops are placed in groups of three by Jason. Which of the below shows the fact ?

   (A) 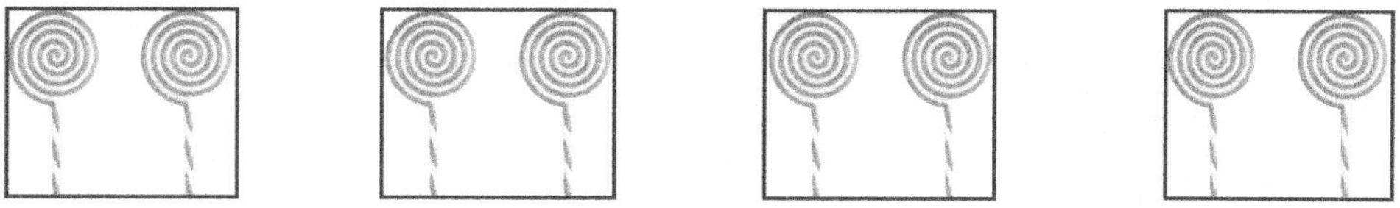

   (B) 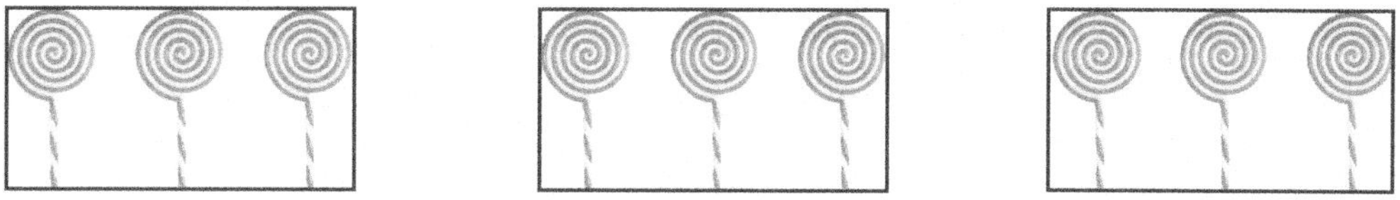

   (C) 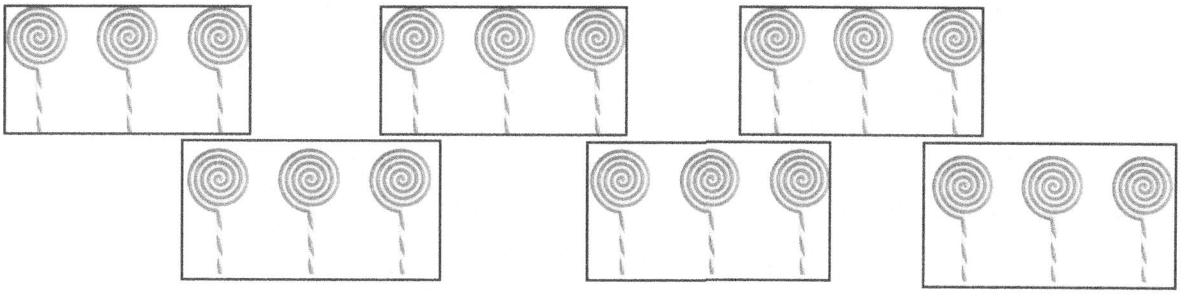

   (D) 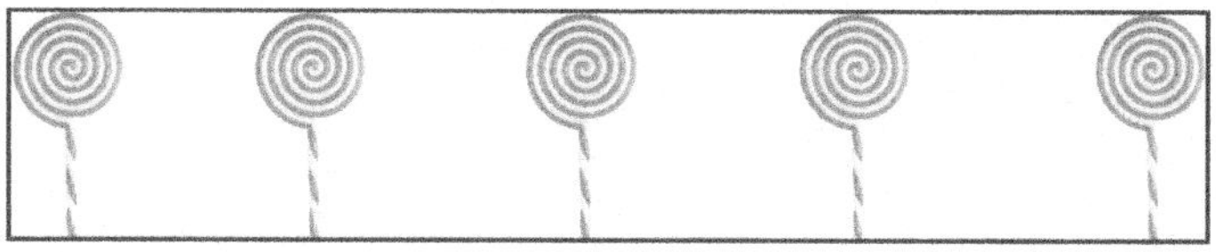

4. Which of the below statement is true ?

(A) $\frac{1}{2} < \frac{1}{4}$ (B) $\frac{1}{2} = \frac{2}{4}$

(C) $\frac{1}{2} > \frac{5}{6}$ (D) $\frac{1}{2} > \frac{3}{4}$

5. Which of the below statement is true ?

(A) 55,505 < 55,005

(B) 55,505 > 55,555

(C) 55,505 = 55,504

(D) 55,505 > 55,500

6. The below figure represents one whole part

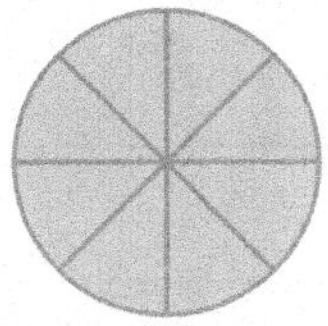

What does the below model represents ?

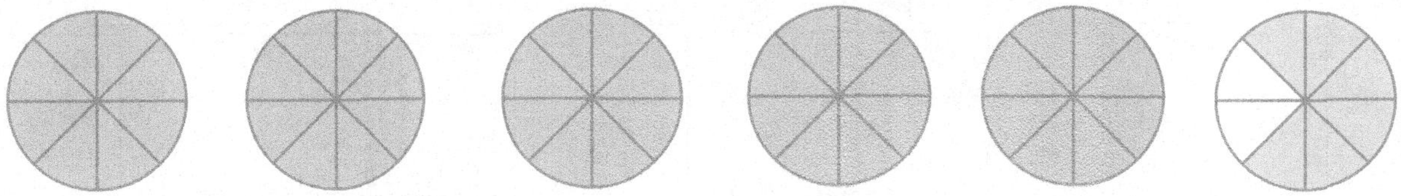

(A) $4\frac{3}{4}$ (B) $5\frac{3}{4}$ (C) $6\frac{3}{4}$ (D) $7\frac{3}{4}$

7. The figure below is shaded to represent the number one.

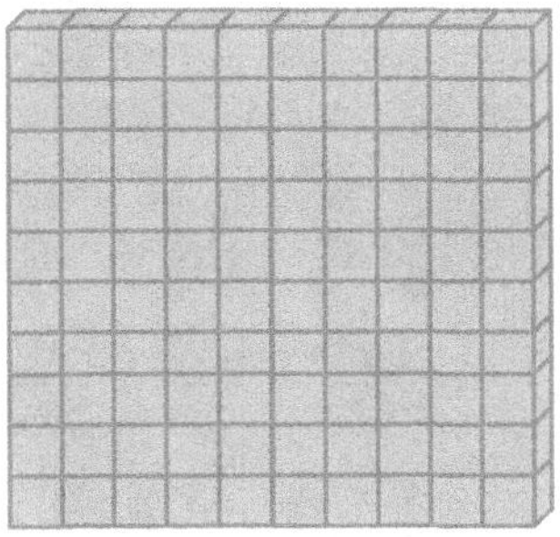

The below model represents which number ?

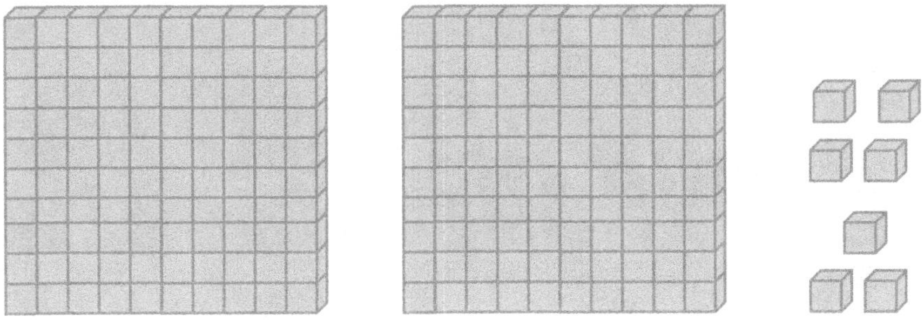

(A) 2.07 (B) 2.6 (C) 2.7 (D) 2.067

8. This model is shaded to show a fraction of a whole.

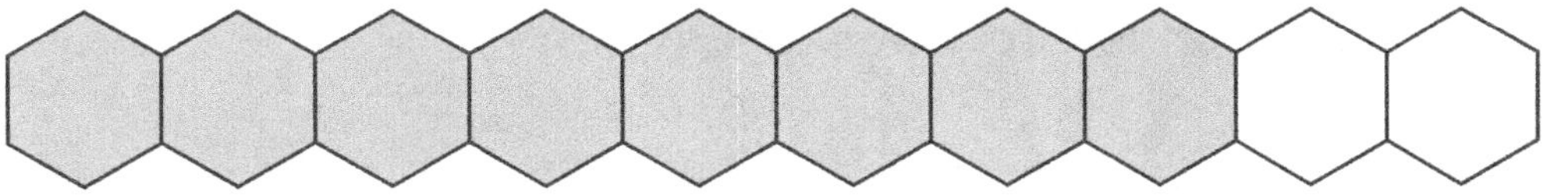

Which of the following is shaded to show a fraction with a value GREATER THAN the one above ?

(A) 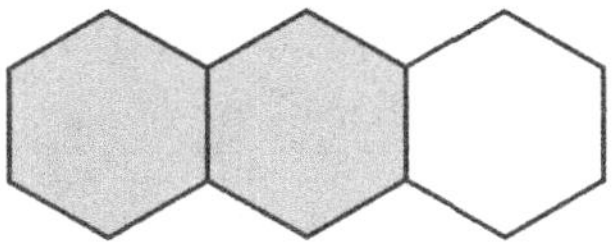

(B) 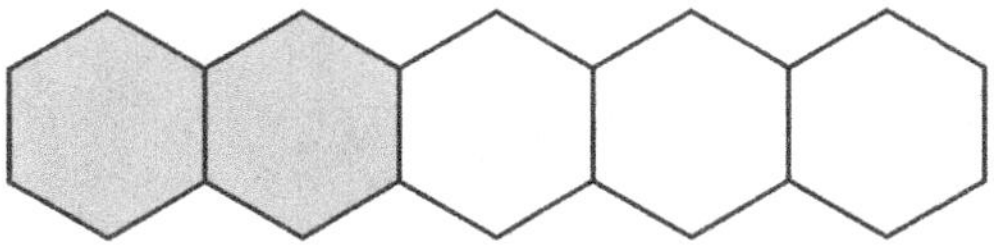

(C) 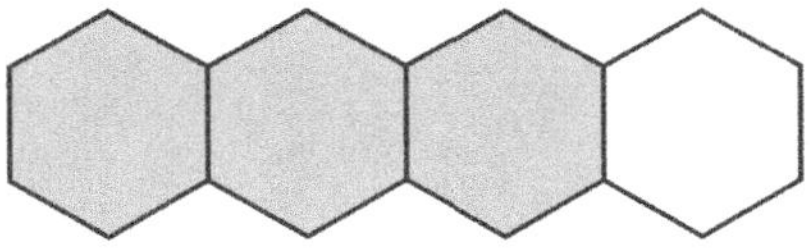

(D) 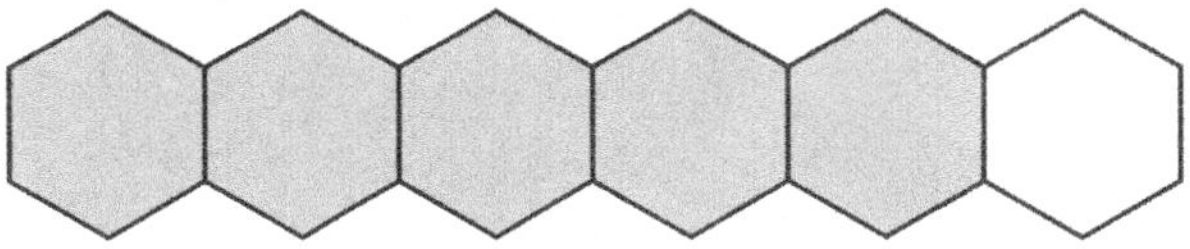

9. Which picture box below represents an even number of triangles ?

(A)

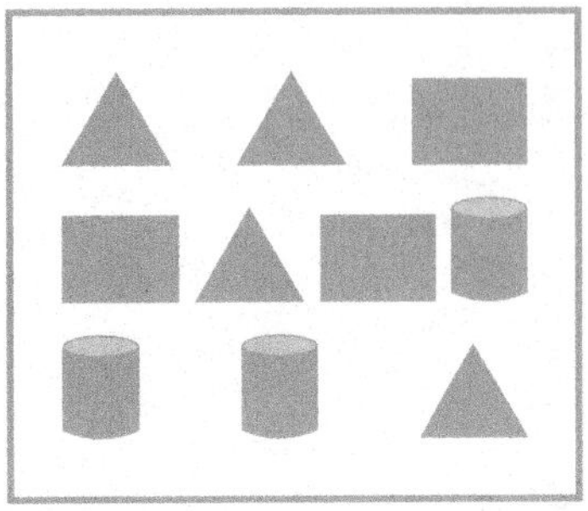

(B)

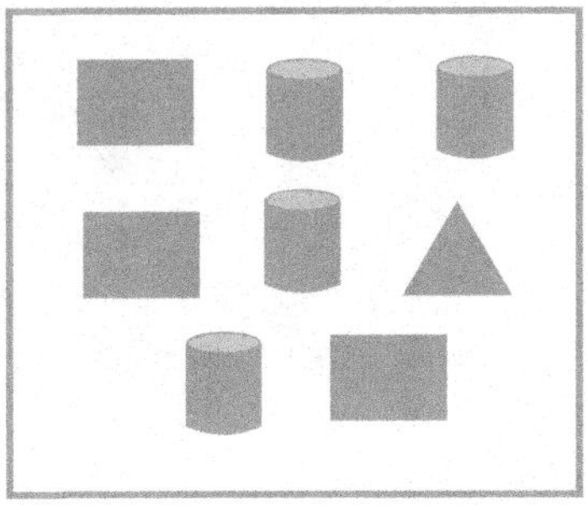

(C)

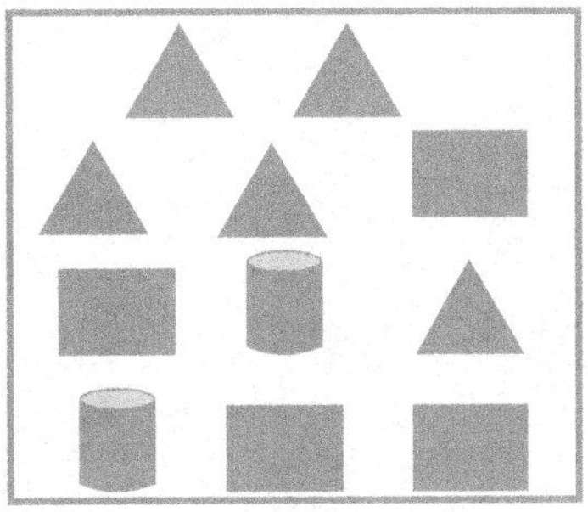

(D)

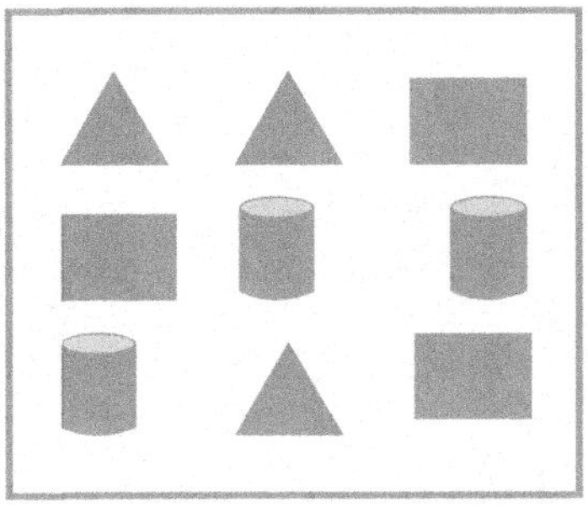

10. What is 777,115 rounded to the nearest ten ?

(A) 777,115

(B) 777,100

(C) 777,120

(D) 770,000

11. Sachi solved 555,805 games in puzzles this year. What is the value of eight in 555,805 ?

(A) 800,000

(B) 8

(C) 80

(D) 800

12. This  represents a value of 5.

Which model represents the value of 2 in 861,259 ?

(A) 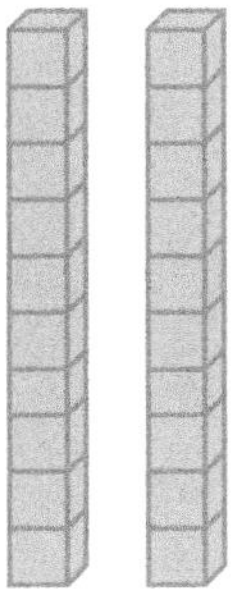

(B) 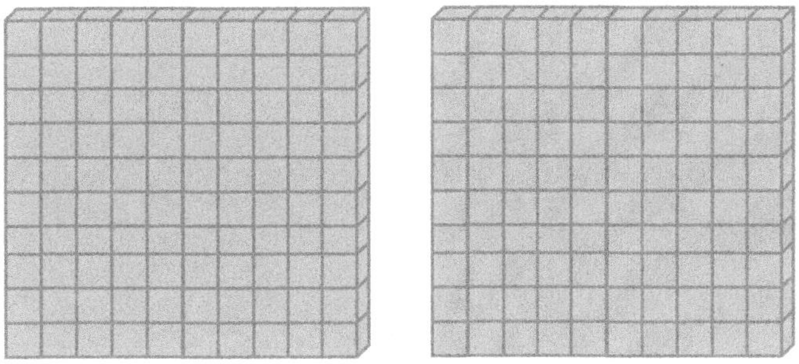

(C) 

(D) 

13. What fraction of Parrots are there ?

(A) $\frac{4}{5}$ (B) $\frac{3}{4}$ (C) $\frac{2}{5}$ (D) $\frac{4}{1}$

14. Evaluate 33 - [?] = 5

Which of the below fact can be used to complete the given number sentence ?

(A) 33 + 5

(B) 33 × 5

(C) 33 ÷ 5

(D) 33 - 5

15. Mary solves 6 math problems in 5 minutes. How many puzzles does she solve 10 minutes ?

(A) 30

(B) 20

(C) 12

(D) 14

16. Evaluate

$20 \div 4 = \boxed{?}$

(A) 3

(B) 5

(C) 6

(D) 9

17. Jason has fifty three Chess cards and thirty nine Math cards

53	39
Chess Cards	Math Cards

How many more chess cards than math cards do Jason have ?

(A) 11

(B) 14

(C) 102

(D) 92

18. Evaluate

$7,608 + 392 = \boxed{?}$

(A) 8,090

(B) 8,100

(C) 8,108

(D) 8,000

19. This model is shaded to show two values.

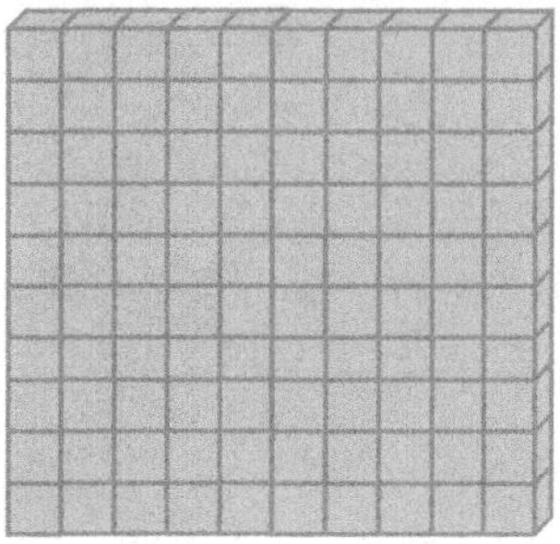

What decimal is shown by the model below ?

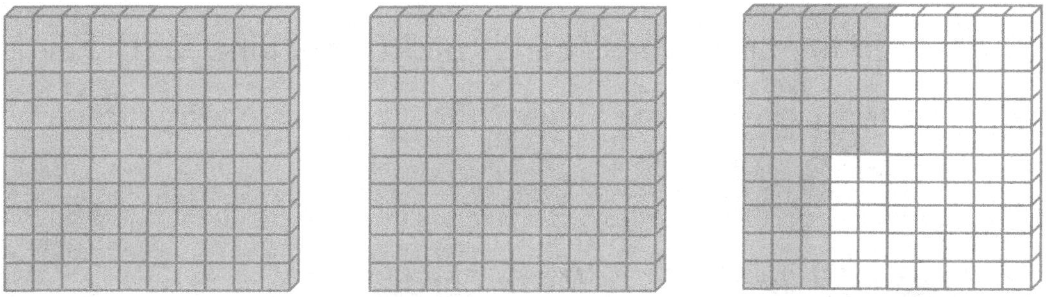

(A) 4.38 (B) 4.44 (C) 4.8 (D) 4.033

20. Evaluate

$$\begin{array}{r} 7\ .\ 05 \\ +\ 2\ .\ 04 \\ \hline \\ \hline \end{array}$$

(A) 9.09

(B) 10.9

(C) 12.3

(D) 25.8

21. This model is shaded to show one whole.

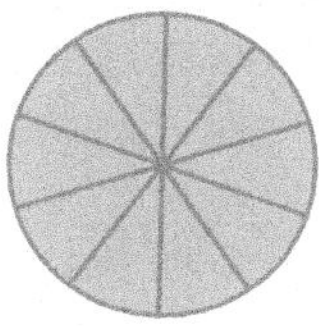

The two models below are each shaded to show a fraction

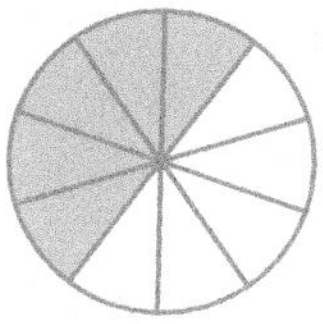

Model 1

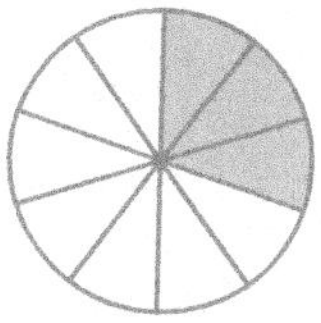

Model 2

One fraction is added from the other. Which model is shaded to show the difference ?

(A) 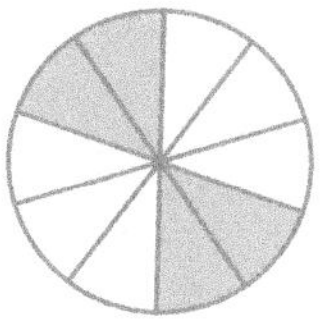

(B) 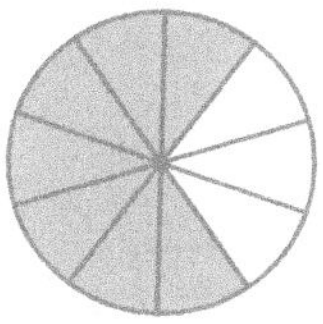

(C) 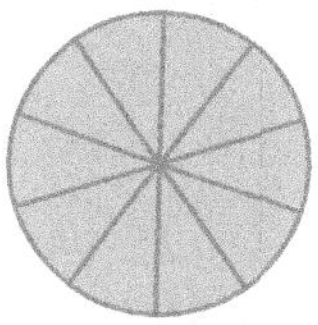

(D) 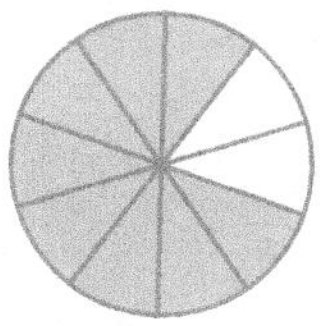

22. Evaluate

9 x 9 = [?]

(A) 71

(B) 81

(C) 90

(D) 91

23. Which best shows $2 \times 5 = \boxed{?}$

(A) 8

(B) 12

(C) 10

(D) 14

24. Evaluate $7.9 + 5.5 = \boxed{?}$

(A) 13 . 4

(B) 12 . 2

(C) 16 . 5

(D) 18 . 3

25. Lucy uses the math fact $\boxed{11 \times 12 = 132}$ to solve a math problem. Which of the below could be the problem she is trying to solve ?

(A) $132 - \square = 12$

(B) $\square \times 11 = 12$

(C) $\square + \square = 11$

(D) $\square \div 12 = 11$

26. Tony's basketball team makes 16 baskets all together to win the game. Tony makes 8 baskets in all. What fractional part of the baskets did tony made ?

(A) $\frac{1}{8}$ (B) $\frac{8}{15}$ (C) $\frac{1}{2}$ (D) 2

27. The number sentence given below models a property of multiplication facts.

$$7 \times 3 = 3 \times 7$$

Which of the following number sentences shows the same property as above ?

(A) $5 \times 3 = 15 \times 1$

(B) $6 \times 9 = 9 \times 6$

(C) $9 \times 4 = 6 \times 6$

(D) $11 \times 11 = 11 + 11$

28. What number goes in the box to make the below number sentence true ?

$$96 - \boxed{?} = 50$$

(A) 146

(B) 75

(C) 46

(D) 45

29. Rosy was born about 6 months ago . What is her approximate age in weeks now ?

(A) 25

(B) 150

(C) 365

(D) 13

30. Identify the closest temperature shown in the below figure ?

(A) 26^{0} F

(B) 150^{0} F

(C) 52^{0} F

(D) 13^{0} F

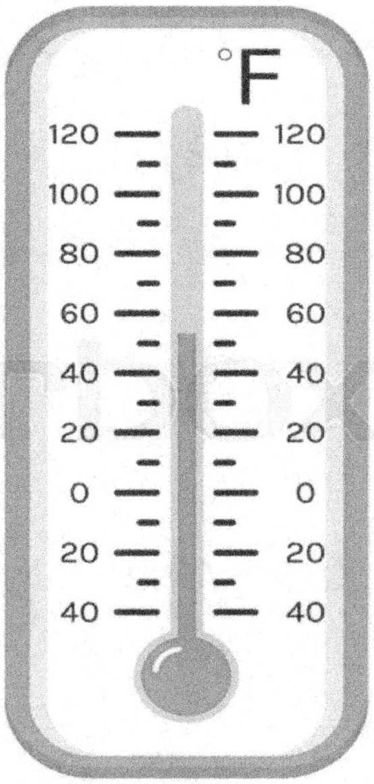

31. Identify the picture below ?

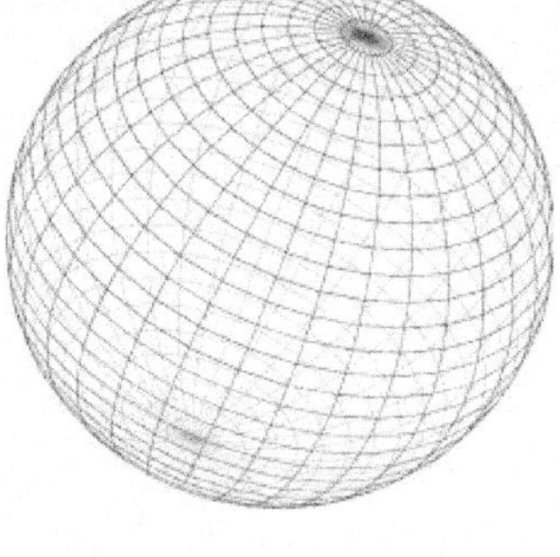

(A) Triangle (B) Sphere (C) Oval (D) 3-D object

32. Length of one tooth pick is equal to 1 cm. What is the closest length of the table ?

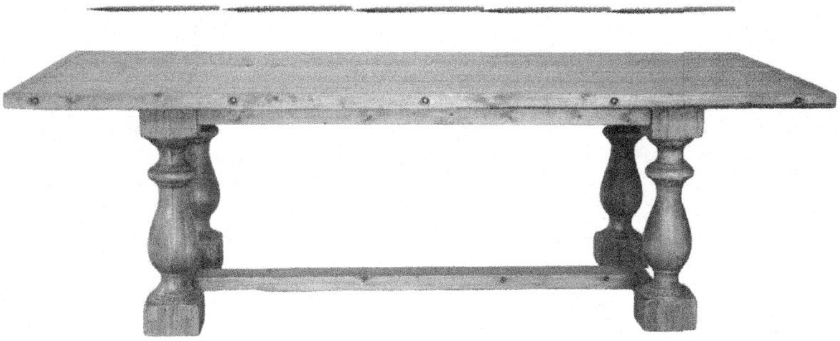

(A) 3.02 cm

(B) 3 cm

(C) 5.2 cm

(D) 10 cm

33. Which part of the picture on the right side represents a square ?

(A) House

(B) Doors

(C) Windows

(D) Room

34. Olivia is wondering which of the below is showing the time of 11.30PM ?

(A) 

(B) 

(C) 

(D) 

35. Identify which of the following figures has the line of symmetry across the dotted line ?

(A) 

(B) 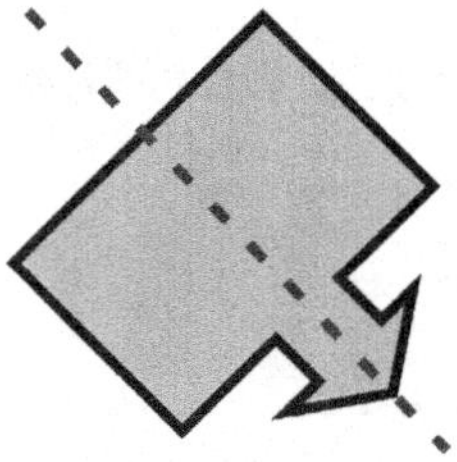

(C) 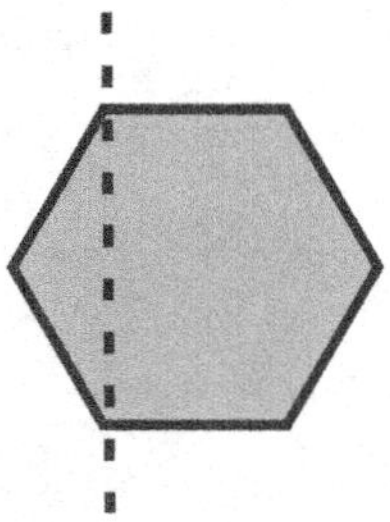

(D) 

36. A Rectangle has ...... ?

(A) Only three sides are equal

(B) All sides are equal

(C) Opposite sides are not parallel

(D) Opposite sides are equal

37. Identify which of the following figures has the line of symmetry across the dotted line ?

(A) 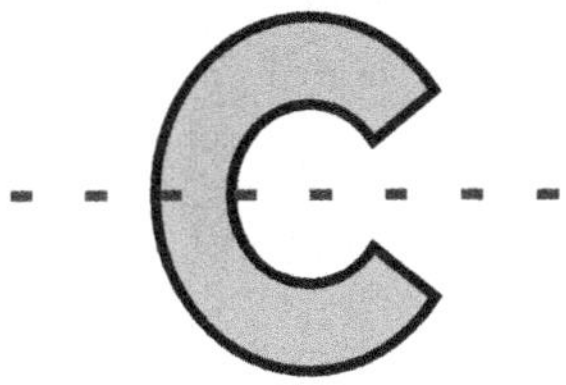

(B) 

(C) 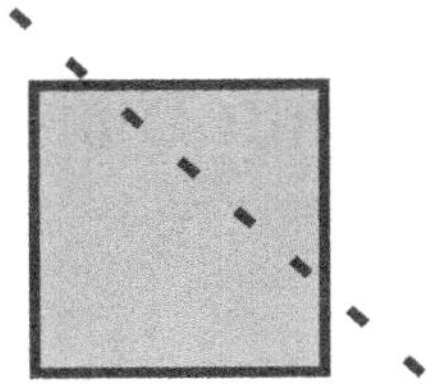

(D) 

38. Below data shows the number of cars repaired in Zack's garage this quarter ?

Month	October - 2018	November - 2018	December - 2018
Number Repaired	15	6	18

He prepares the below chart representing the above data

Month	Number Repaired
October - 2018	
November - 2018	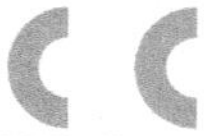
December - 2018	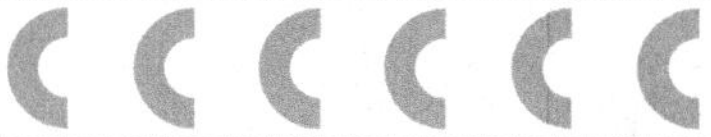

The symbol  represents the value of ____________ ?

(A) 2 (B) 1 (C) 3 (D) 4

39. What is the perimeter of the below rectangle ABCD in inches ?

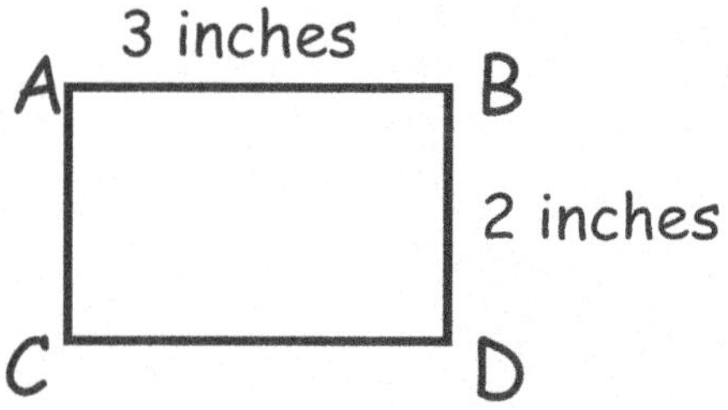

(A) 8 inches (B) 11 inches (C) 9inches (D) 10 inches

40. Andy has

in savings this week.

He spent for his lunch

on this Saturday.

How much money does he have now ?

(A) $ 7 (B) $ 1.30

(C) $ 3.35 (D) $ 1.10

41. Sandy collected below leafs for her school project.
Which of the below bar graph correctly shows the number of each kind of leafs she collected ?

Which bar graph correctly shows the number of each kind of leaf ?

(A)

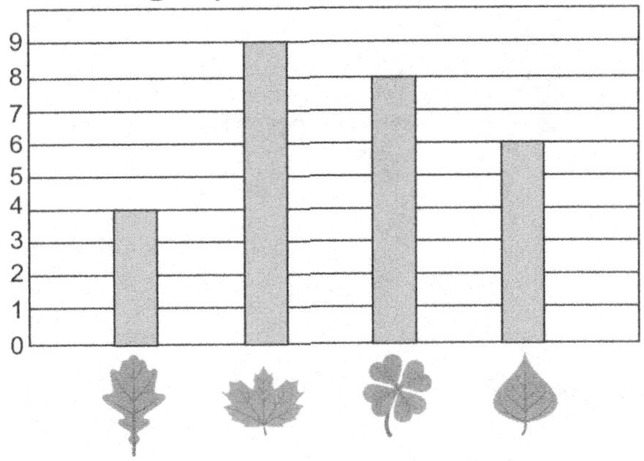

(B)

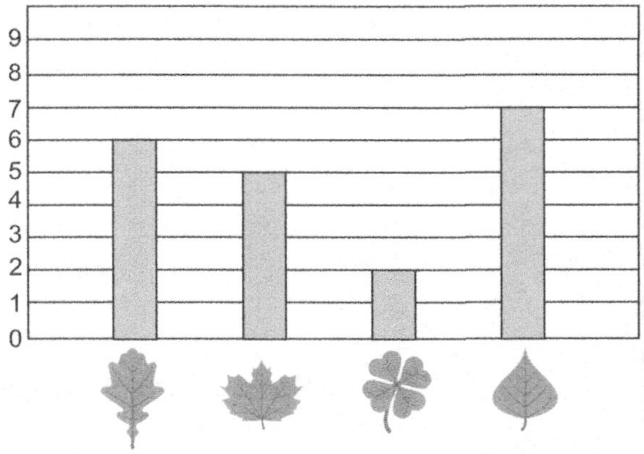

(C)

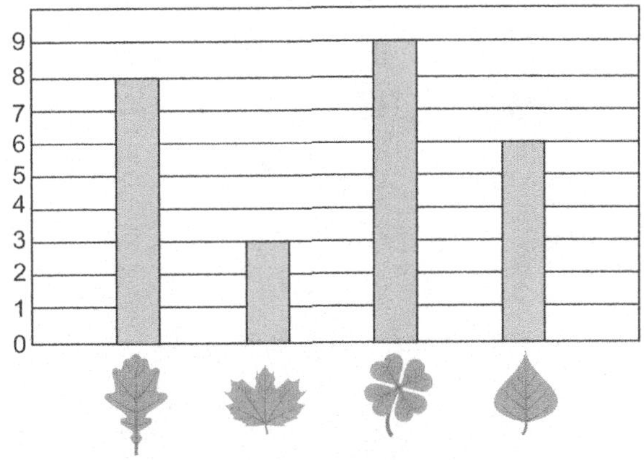

(D)

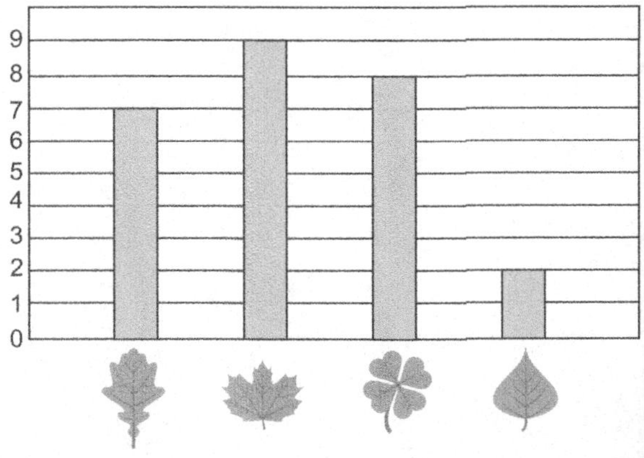

42. Below numbers are written on individual cards and are placed in a jar after carefully placing each of them in envelops. Rik picks an envelope from the jar what is the chance of the number in the envelop being a two digit number ?

22 12 88 35 77

(A) Certain

(B) Likely, but not certain

(C) Unlikely, but not impossible

(D) Impossible

43. Grade 3 of Wise school collects story books for six weeks for a book donation project. Below graph shows the weeks horizontally.
What is missing vertically ?

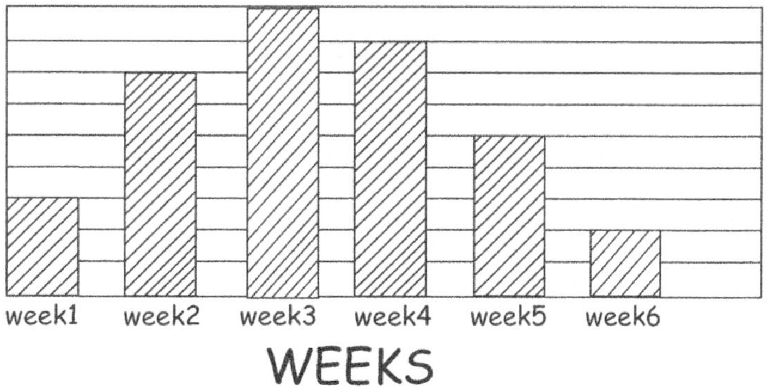

(A) Total Stamps

(B) Chocolates eaten

(C) Number of story books

(D) Average story books collected

44. Cathy has two types of beads to make jewellery. She sorted her beads into the following two groups. She placed beads of the same shape and shade into a group.

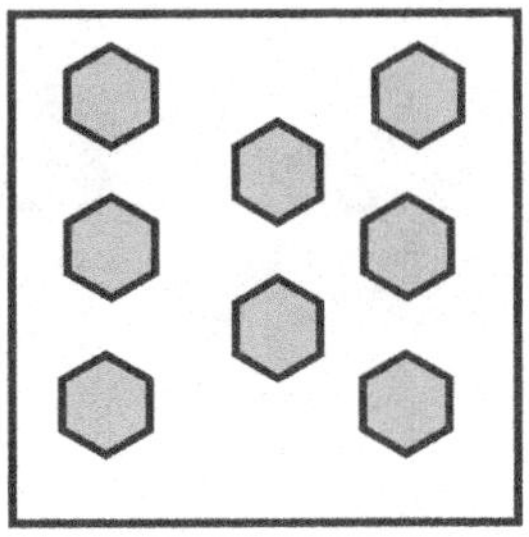

Group B

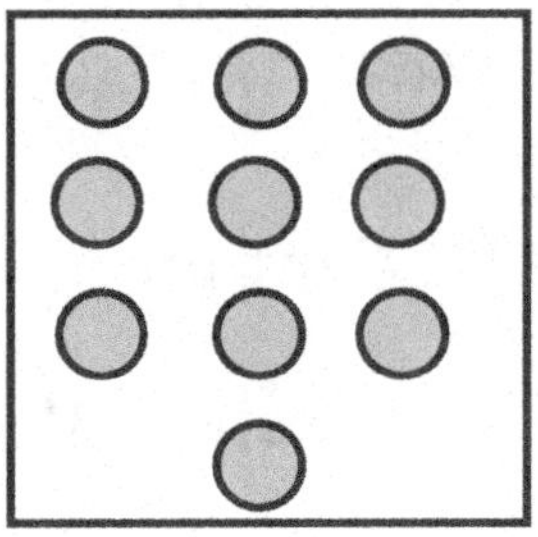

Group R

Which shape should Cathy include in Group R ?

(A)  (B)  (C)  (D) None

45. Find the next two numbers in the series ?

122 , 119 , 116 , 113 ,.....?

(A) 103, 105 (B) 111, 109

(C) 110, 107 (D) 105, 101

46. Which number can be placed on the line to make the number sentence true ?

101 X [?] = 505

(A) 50 (B) 5 (C) 11 (D) 404

47. Using the skip counting rule what are the next three numbers in the series ?

300 , 307 , 314 , 321 , 328 ,... ?

(A) 328 , 329 , 336 (B) 334 , 341 , 348

(C) 335 , 356 , 355 (D) 335 , 342 , 349

48. The table shows the total costs of different Art pieces that Dora is selling in an art exhibition.

Selling price of Art pieces

Number of Art Pieces	Total Cost
1	$ 11.55
2	$ 23.10
3	$ 34.65
4	$ 46.20
7	$?

Based on the pattern in the table, what will be the total cost for seven Art pieces ?

(A) 71.70 (B) 80.85 (C) 69.75 (D) 79.085

49. Daisy has collected leafs from different trees . She counts them after sorting. Which chart below shows the correct count ?

(A)

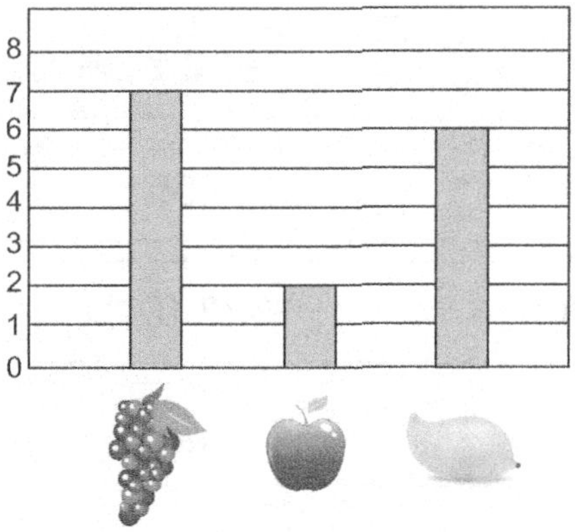

(B)

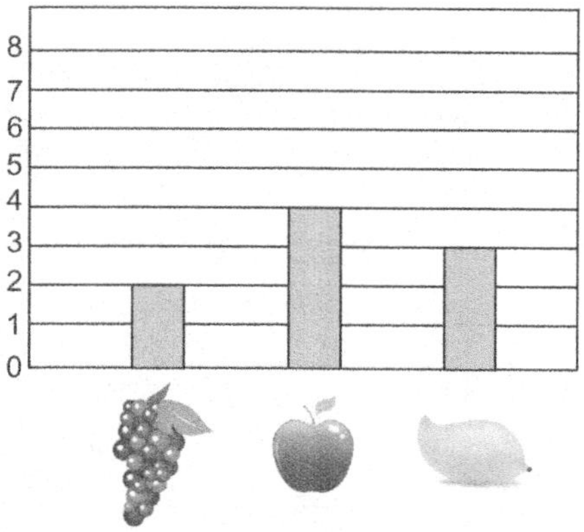

(C)

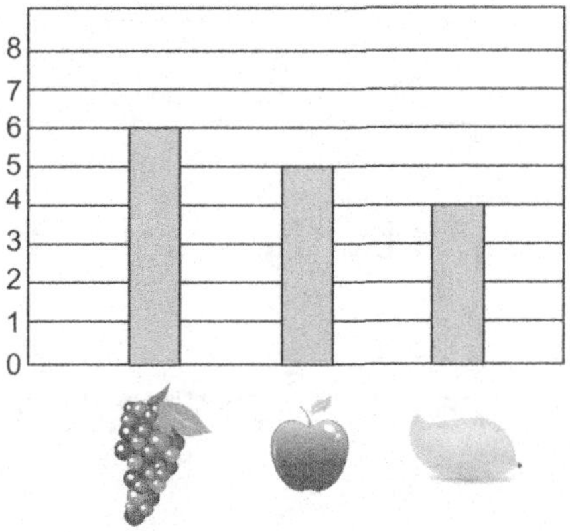

(D)

50. Jack is grouping his twelve marbles as below to represent a multiplication fact. Which of the below models the multiplication fact ?

(A) 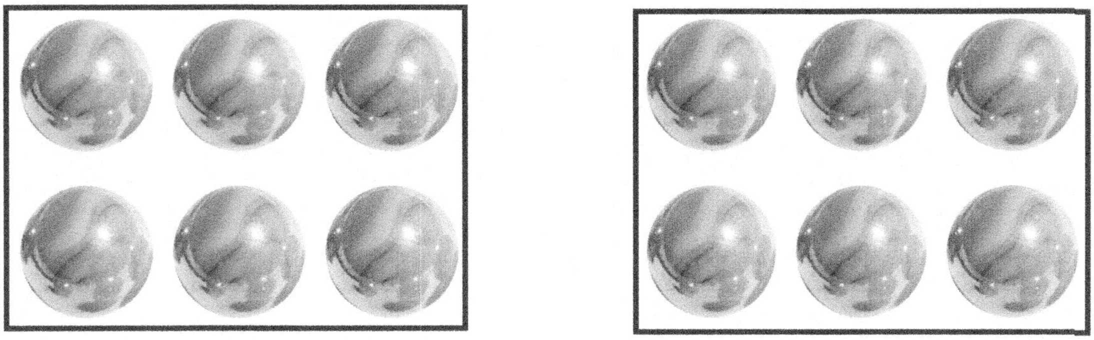

(B) 

(C) 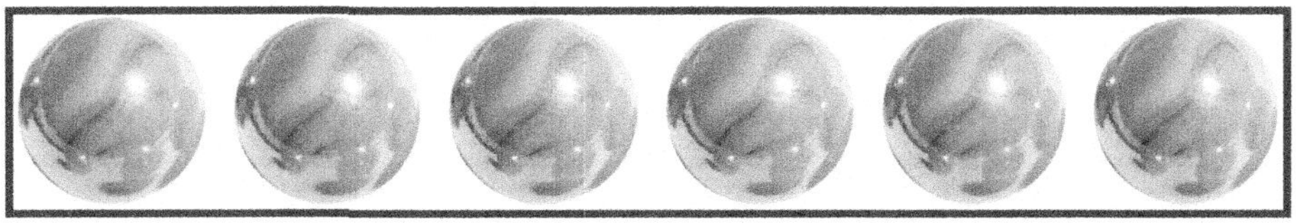

(D) 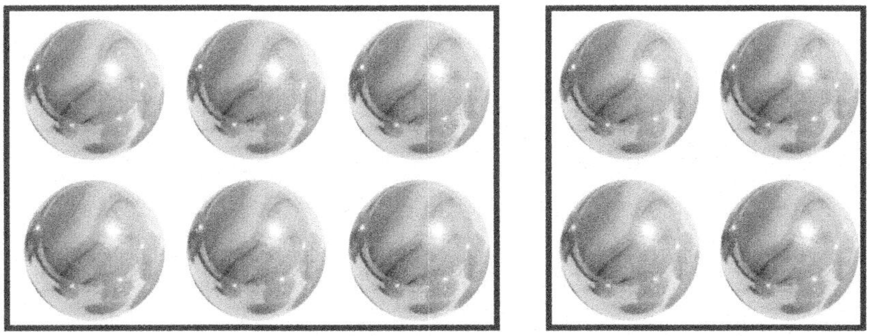

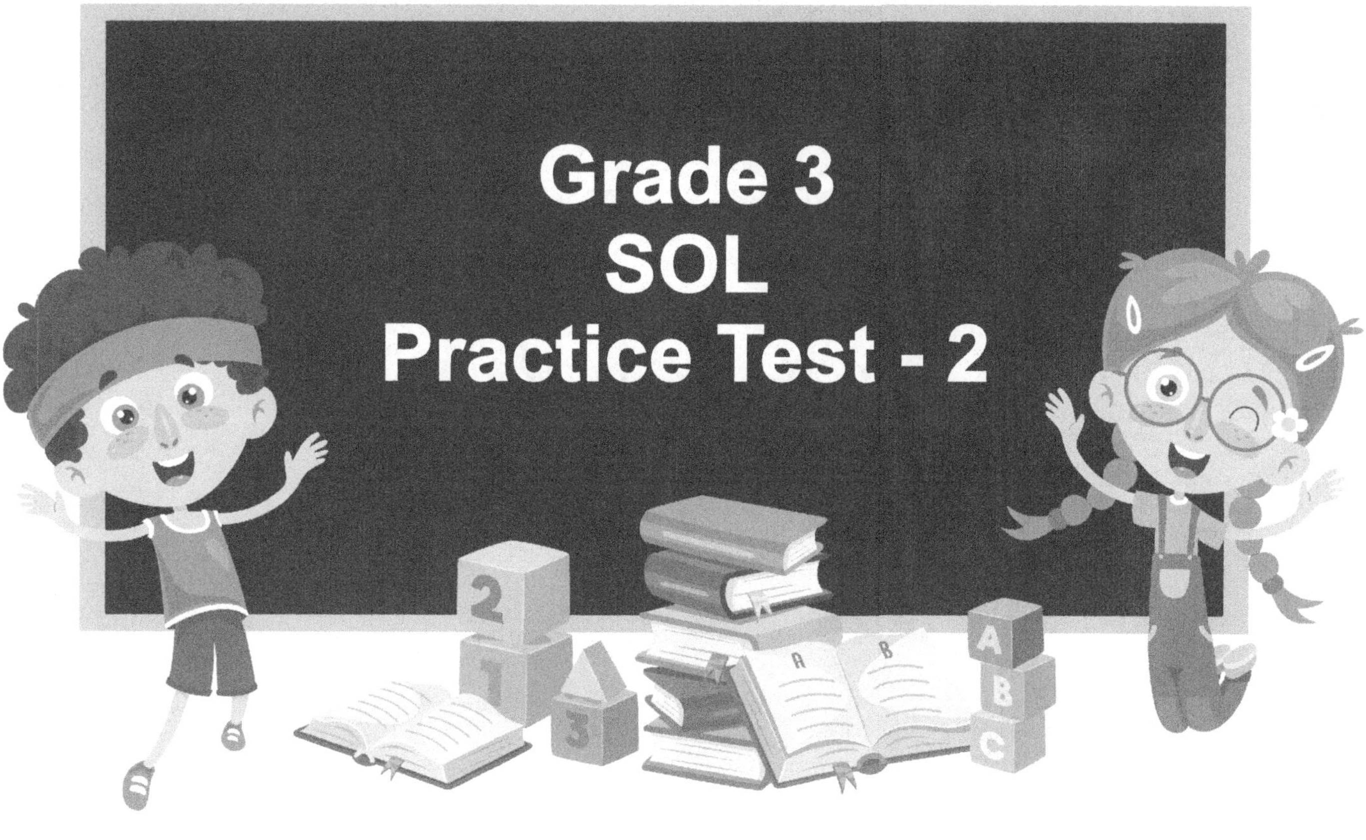
Grade 3
SOL
Practice Test - 2

1. The table shows the total costs of different numbers of Stamps.

Cost of Stamps

Number of Stamps	Total Cost
1	$ 0.30
2	$ 0.60
3	$ 0.90
4	$ 1.20
11	$?

Based on the pattern in the table, what will be the total cost for 11 Stamps ?

(A) $3.30 (B) $2.03 (C) $1.50 (D) $2.10

2. Using the skip counting rule what is the next number in the series ?

662 , 664 , 666 , 668 ,.... ?

(A) 702 (B) 672 (C) 704 (D) 670

3. Find the next two numbers in the series ?

999 , 888, 777, 666,..... ?

(A) 555 , 333 (B) 555 , 444

(C) 666 , 555 (D) 505 , 444

4. Julia groups fifteen of her hair clips in groups of five. Which of the below shows the fact ?

(A) 

(B) 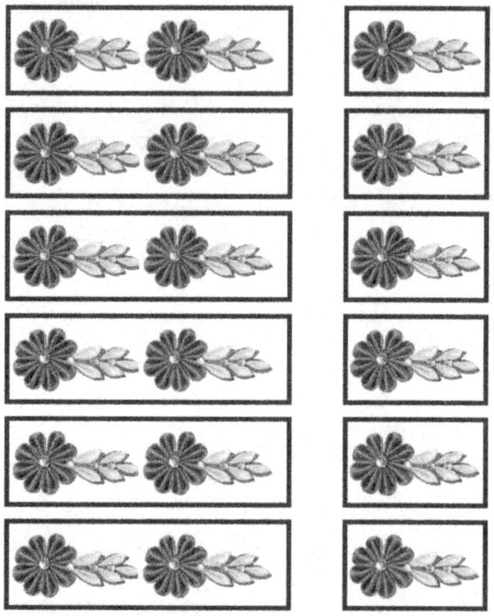

(C) 

(D) 

5. Which number can be placed on the line to make the number sentence true ?

$19 \times \boxed{?} = 57$

(A) 3

(B) 2

(C) 4

(D) 1

6. Pat has sorted her numbers into the following two groups.

Group Peck	Group Tek
2 4 12 202	999 1001
198 5006 998	87 5 2005
768 542 34	65 147 555

Which number should Pat include in group Tek ?

(A) 8608 (B) 50000 (C) 2007 (D) 1998

7. Esha uses the math fact $\boxed{5 \times 7 = 35}$ to solve a math problem. Which of the below could be the problem she is trying to solve ?

(A) $\square \div 5 = 7$

(B) $\square \times 7 = 5$

(C) $\square + \square = 7$

(D) $35 - \square = 5$

8. Below figures are drawn on individual cards and are placed in a jar after carefully placing each of them in envelops. Rik picks an envelope from the jar what is the chance of the figure on the card being a square ?

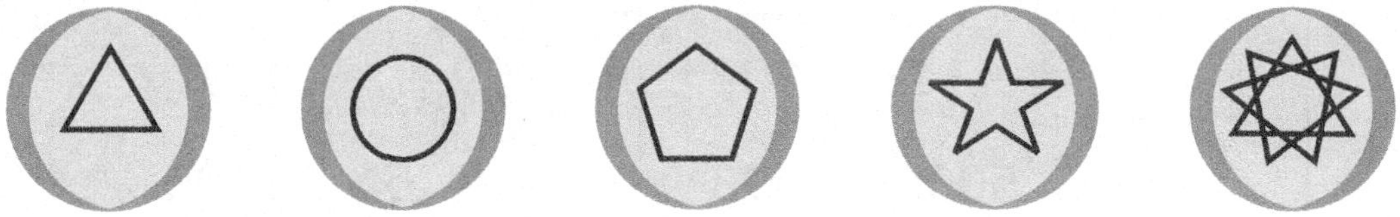

(A) Certain
(B) Likely, but not certain
(C) Unlikely, but not impossible
(D) Impossible

9. Which of the below statement is true ?

(A) $\frac{3}{11} < \frac{4}{15}$
(B) $\frac{3}{13} > \frac{3}{7}$
(C) $\frac{5}{9} < \frac{7}{6}$
(D) $\frac{2}{19} > \frac{5}{7}$

10. These are different toy vehicles Jacob collected for school

Which bar graph correctly shows the number of each kind of toy vehicles ?

(A)

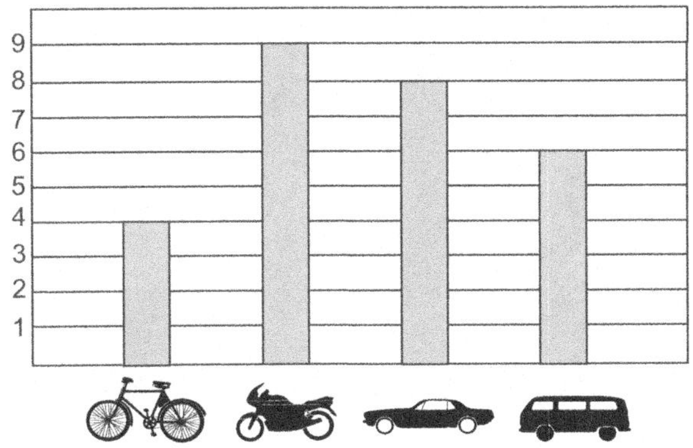

(B)

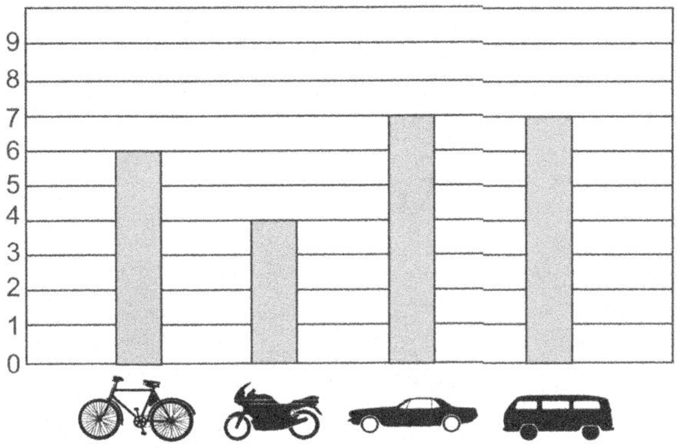

(C)

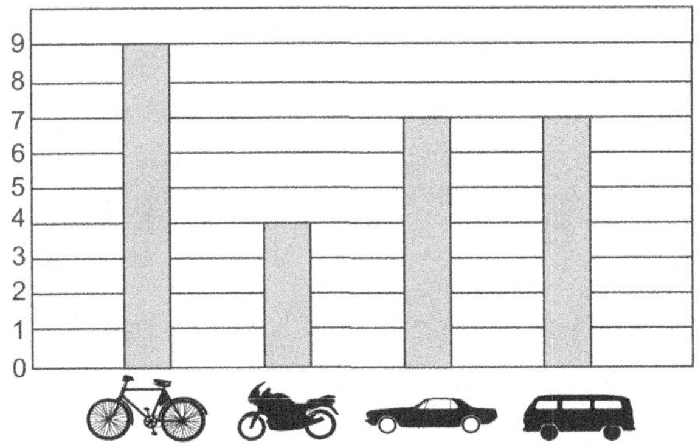

(D)

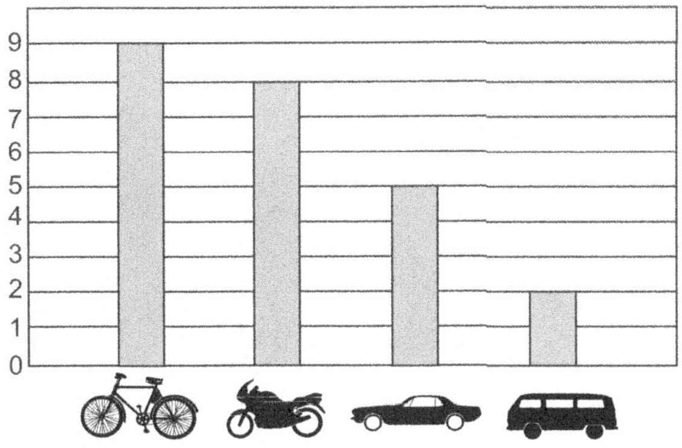

11. Below data shows the number of puzzles sold in Tony's toy store in the last three months ?

Month	October - 2018	November - 2018	December - 2018
Number Repaired	25	35	50

He prepares the below chart representing the above data

Month	Number Repaired
October - 2018	
November - 2018	
December - 2018	

The symbol  represents ?

(A) 5 (B) 2 (C) 10 (D) 3

12. What is the perimeter to the distance around this figure in inches ? (Approximately using the ruler)

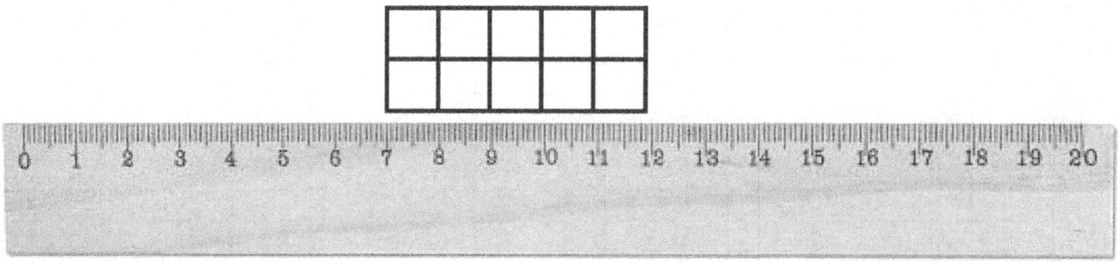

(A) 15 units (B) 14 units (C) 9 units (D) 7 units

13. Jack earned by snow shoveling and received the below amount this week.

After work he went to a movie and spent the below amount. How much money does he have now ?

(A) \$ 2.29 (B) \$ 2.45 (C) \$ 2.20 (D) \$ 2.56

14. A Circle has ......?

(A) Opposite sides are equal
(B) Many sides
(C) Only one Center
(D) 270^0

15. Identify which of the following figures has the line of symmetry across the dotted line ?

(A) 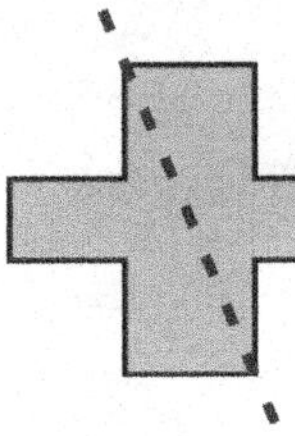

(B) 

(C) 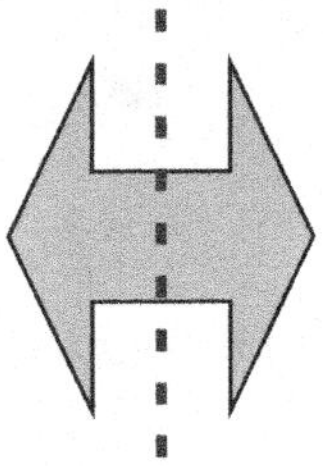

(D) 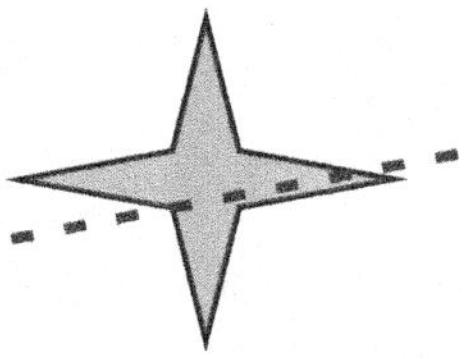

16. Anne is wondering which of the below is showing the time of 2.25 PM ?

(A) 

(B) 

(C) 

(D) 

17. Which part of the below picture represents a circle ?

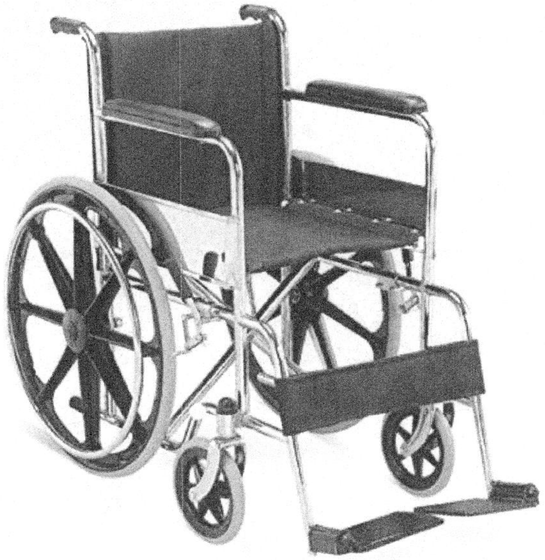

(A) Chair

(B) Arm rest

(C) Wheels

(D) Foot rest

18. What is the closest to the length of book ?

Scale :
- = 1 cm

(A) 8 cm (B) 11 cm (C) 5 cm (D) 19 cm

19. Identify the picture below ?

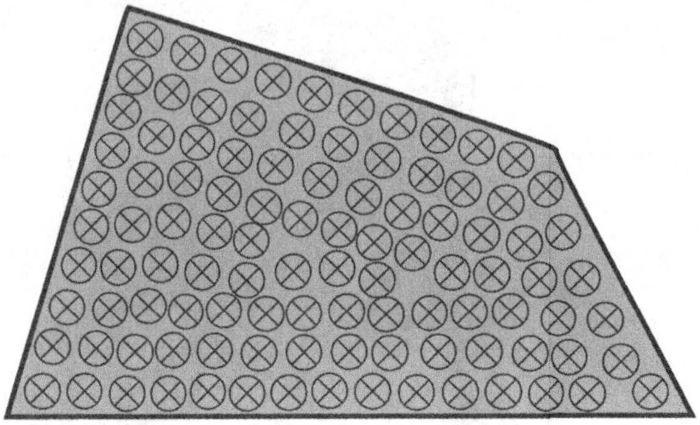

(A) Trapezium (B) Square

(C) Triangle (D) Parallelogram

20. Cory rides his bike for 60 minutes. How many hours did he rode his bike today ?

(A) 23 Hours (B) 60 Seconds

(C) 84 Seconds (D) 1 Hour

21. Identify the closest temperature shown in the figure ?

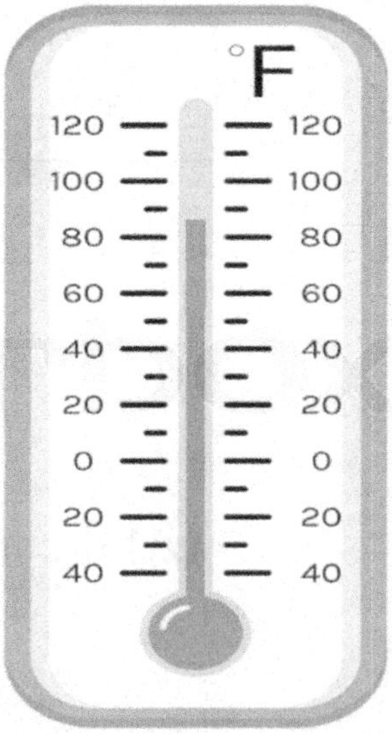

(A) 84^0 Fahrenheit (B) 50^0 Fahrenheit

(C) 10^0 Fahrenheit (D) 100^0 Fahrenheit

22. Length of the running track is measured in which units ?

(A) Cups

(B) Inches

(C) Miles

(D) Centi metres

23. Lucy bakes a cake and cuts it into 10 equal pieces. She gave three pieces to her mom, three to her dad and ate one piece herself. She saved rest of the pie for next day dessert after dinner. What fractional part of the cake was left for dinner ?

(A) $\frac{1}{2}$ (B) $\frac{3}{10}$ (C) $\frac{1}{5}$ (D) $\frac{7}{10}$

24. Which of the figures below appears to have a line of symmetry ?

(A) 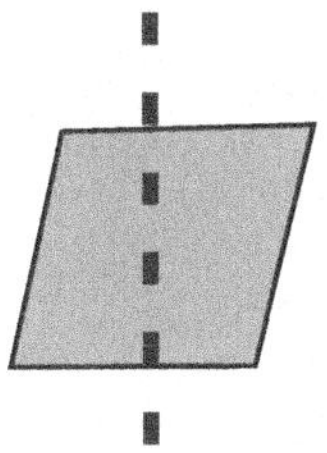

(B) 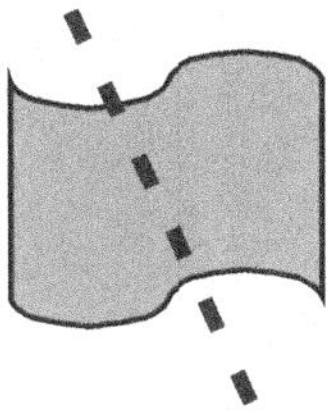

(C) 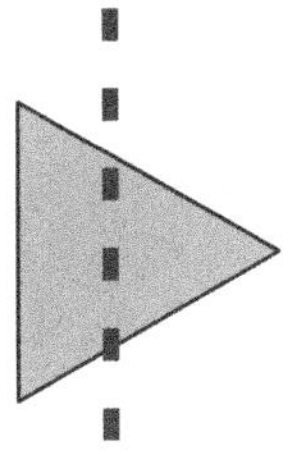

(D) 

25. Evaluate $9.1 + 1.9 = \boxed{?}$

(A) 11.0 (B) 11.10 (C) 10.01 (D)11.01

26. Evaluate $3 \times 7 = \boxed{?}$

(A) 27 (B) 28 (C) 21 (D) 29

27. This model is shaded to show 1 whole.

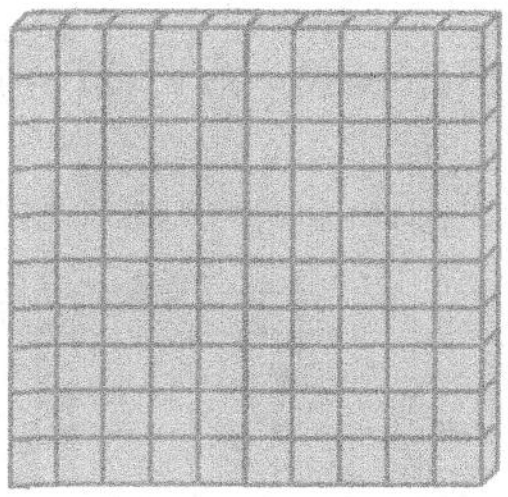

What decimal is shown by the model below ?

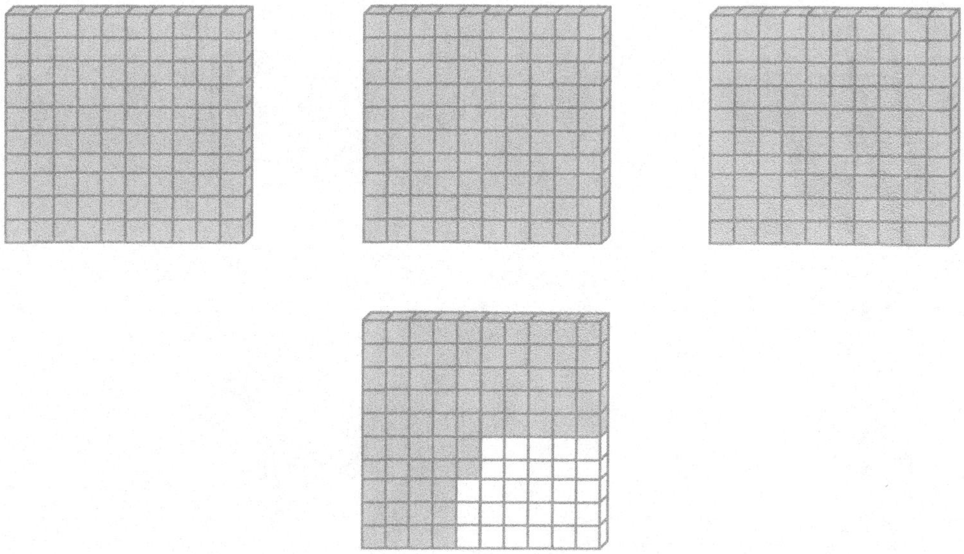

(A) 3.92 (B) 3.72 (C) 3.072 (D) 3.93

28. Which of the below statement is true ?

(A) 101 is equal to 100

(B) 5512 is less than 1900

(C) 838 is greater than 939

(D) 4250 is less than 4299

29. What fraction of rose flowers are there from the below group of roses and leafs ?

(A) $\frac{3}{11}$

(B) $\frac{5}{11}$

(C) $\frac{1}{11}$

(D) $\frac{5}{6}$

30. If this  represents a value of 1.Which model represents the value of 5 in 84,590 ?

(A)

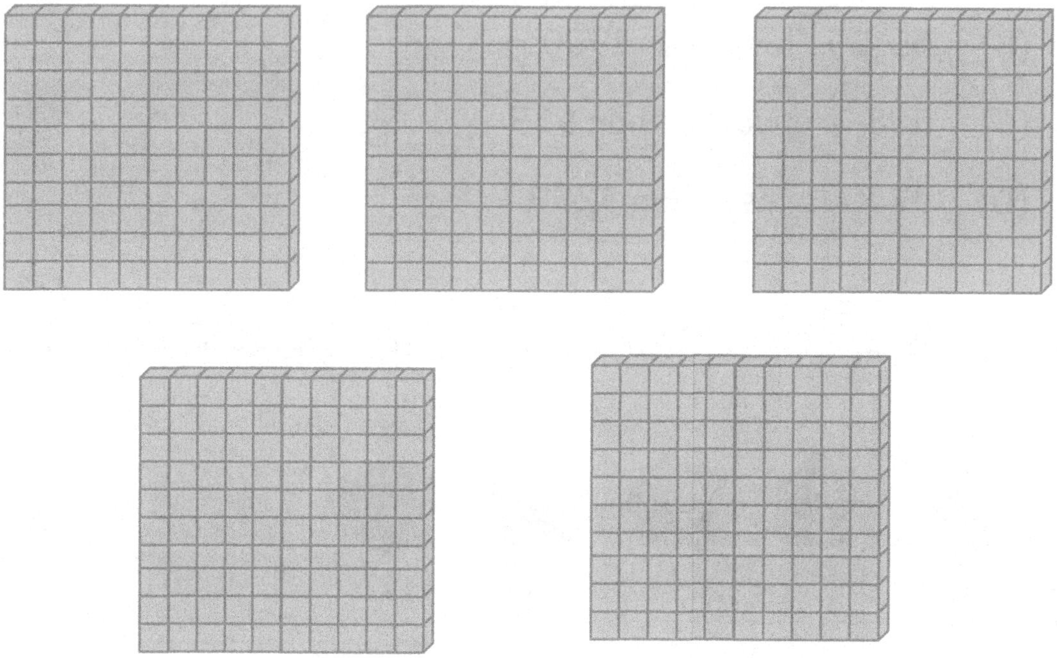

(B)

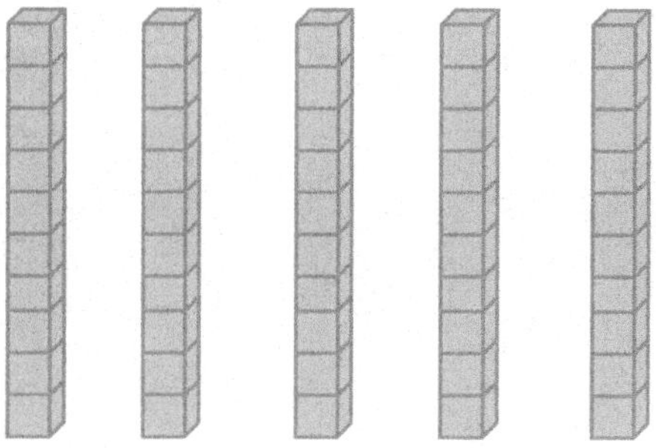

(C)

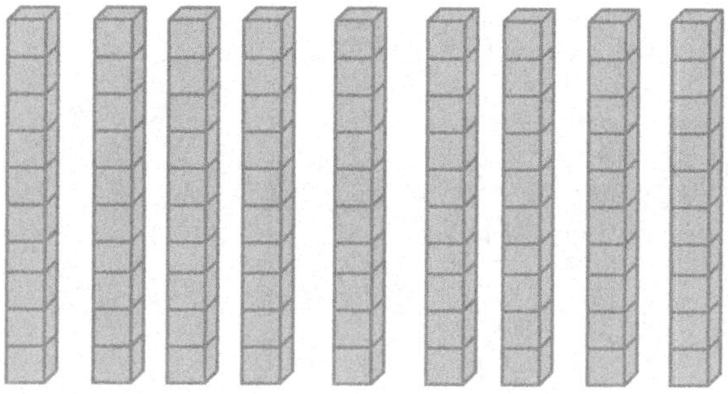

(D)

31. Saachi solved 708,059 puzzles this year.
What is the value of nine in 708,059 ?

(A) 70000

(B) 9

(C) 90

(D) 900

32. The figure below is shaded to represent the number 1.

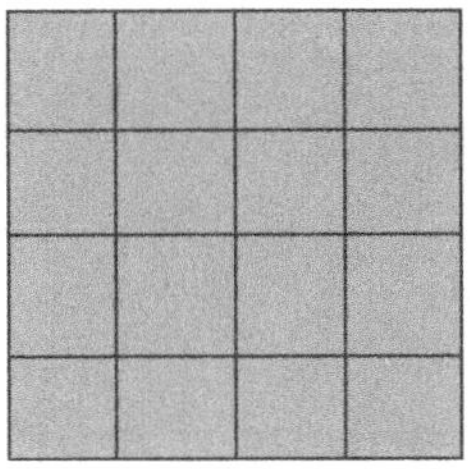

The below model represents which number ?

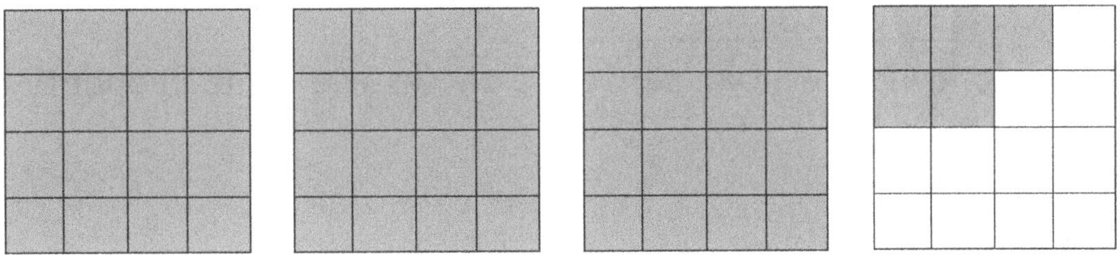

(A) $\frac{5}{16}$

(B) $1\frac{11}{16}$

(C) $4\ \frac{5}{8}$

(D) $3\ \frac{5}{16}$

33. This model is shaded to show a fraction of a whole.

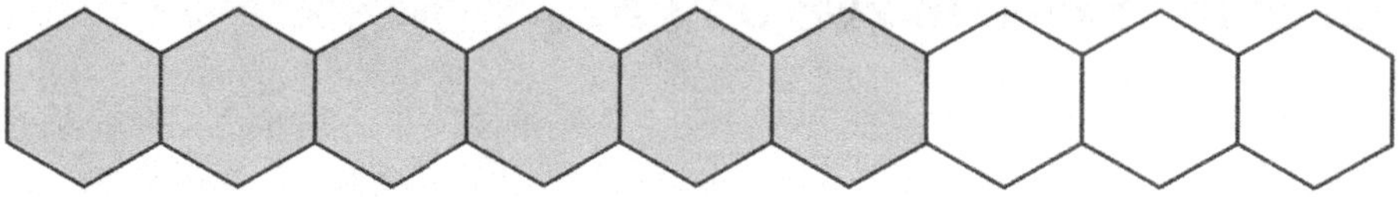

Which of the following is shaded to show a fraction with a value EQUAL the one above ?

(A) 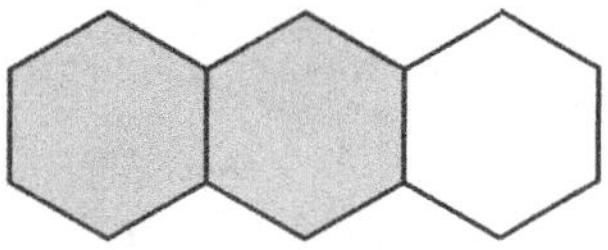

(B) 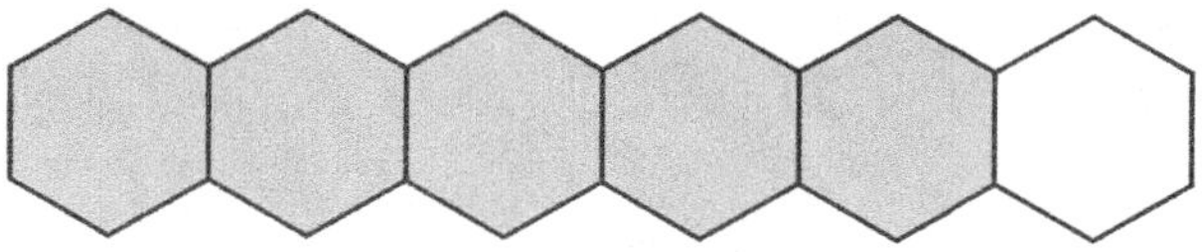

(C) 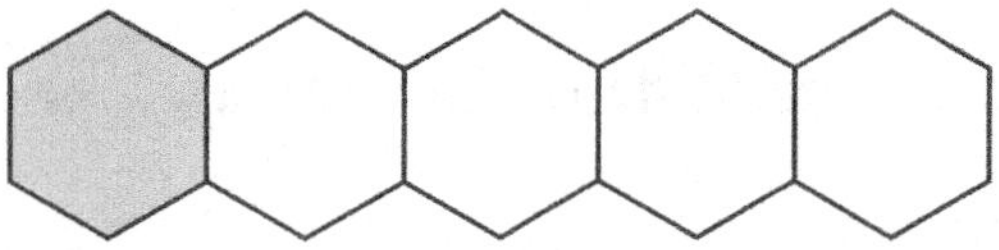

(D) 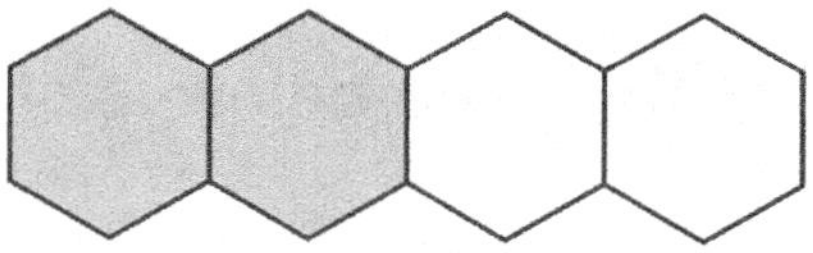

34. The below number sentence shows a multiplication property.

25 × 4 = 4 × 25

Which of the following number sentence shows the same property as above ?

(A) 21 × 3 = 3 × 21

(B) 13 × 5 = 65

(C) 5 × 6 = 16 × 4

(D) 41 × 1 = 41

35. What is 5,875 rounded to the nearest ten ?

(A) 5,900

(B) 5,880

(C) 5,870

(D) 6,000

36. What is 7,956,732 nearest to ten thousand ?

(A) 7,960,000

(B) 7,960

(C) 7,950,000

(D) 7,957,000

37. Rik solves 6 puzzles in a minute. How many puzzles does he solve in 6 minutes ?

(A) 24

(B) 13

(C) 10

(D) 36

38. Evaluate $100 \div 4 = \boxed{?}$

(A) 10 (B) 25 (C) 40 (D) 5

39. Molly has 29 hand bags and 16 pairs of shoes
How many more hand bags does she have than shoe pairs ?

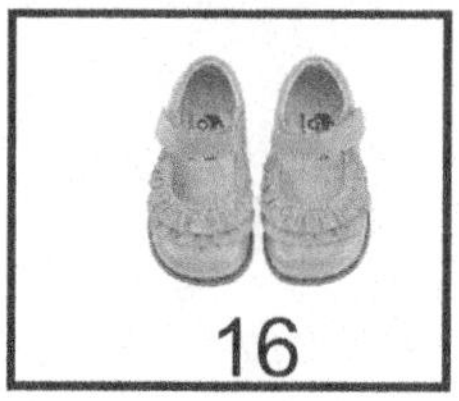

(A) 45 (B) 23 (C) 13 (D) 18

40. Evaluate

$$6001 + 3112 = \boxed{?}$$

(A) 9113

(B) 9200

(C) 9314

(D) 9650

41. Evaluate

$$\begin{array}{r} 5.05 \\ -\ 3.14 \\ \hline \\ \hline \end{array}$$

(A) 8.19

(B) 1.91

(C) 1.81

(D) 7.19

42. What number needs to go in the empty box to make the below number sentence true ?

$$94 + \square = 61$$

(A) 155 (B) 28 (C) 30 (D) 33

43. This model is shaded to show one whole.

The two models below are each shaded to show a fraction.

One fraction is subtracted from the other. Which model is shaded to show the difference ?

(A) $\frac{9}{8}$

(B) $1\frac{1}{8}$

(C) $\frac{3}{8}$

(D) $1\frac{4}{5}$

44. Which of the below shows "Four hundred four thousand, four hundred four" in standard form ?

(A) 4,000,444

(B) 400,414

(C) 404,044

(D) 404,404

45. Which of the below is true ?

(A) 10122 > 10021

(B) 10123 < 10122

(C) 10052 > 10551

(D) 10371 < 10370

46. Which number can be placed on the line to make the number sentence true ?

777 X [?] = 777

(A) 0 (B) 1 (C) 2 (D) 3

47. Cool music club takes voting from students on various music instrument they like to play. The data is plotted as below.

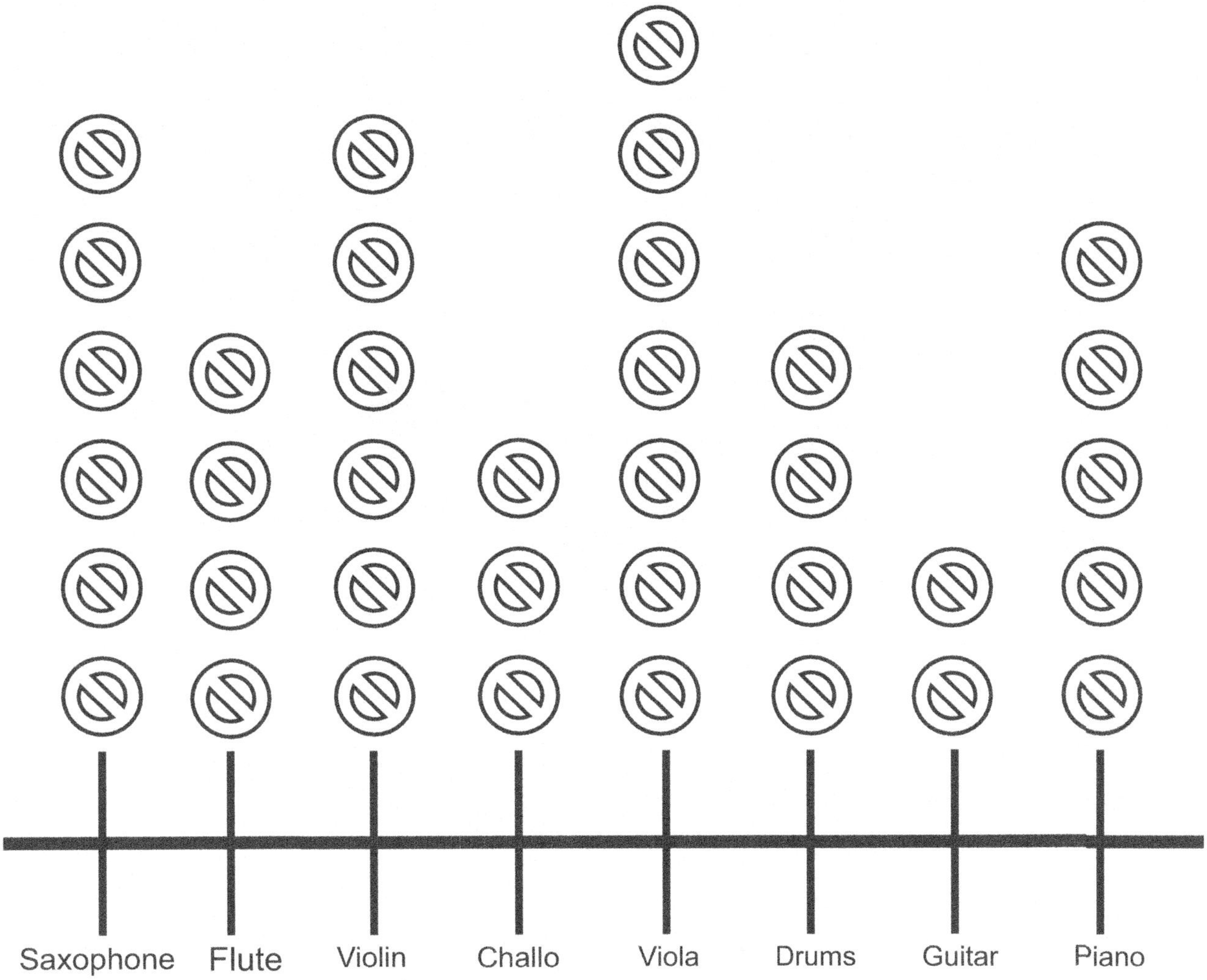

Cool Music Club

What is the most popular music instrument ?

(A) Saxophone (B) Drums

(C) Violin (D) Viola

48. Alice colored small squares in the shape of an alphabet from the below grid. If □ is equal to one square centimeters. What is the area of the alphabet "V" ?

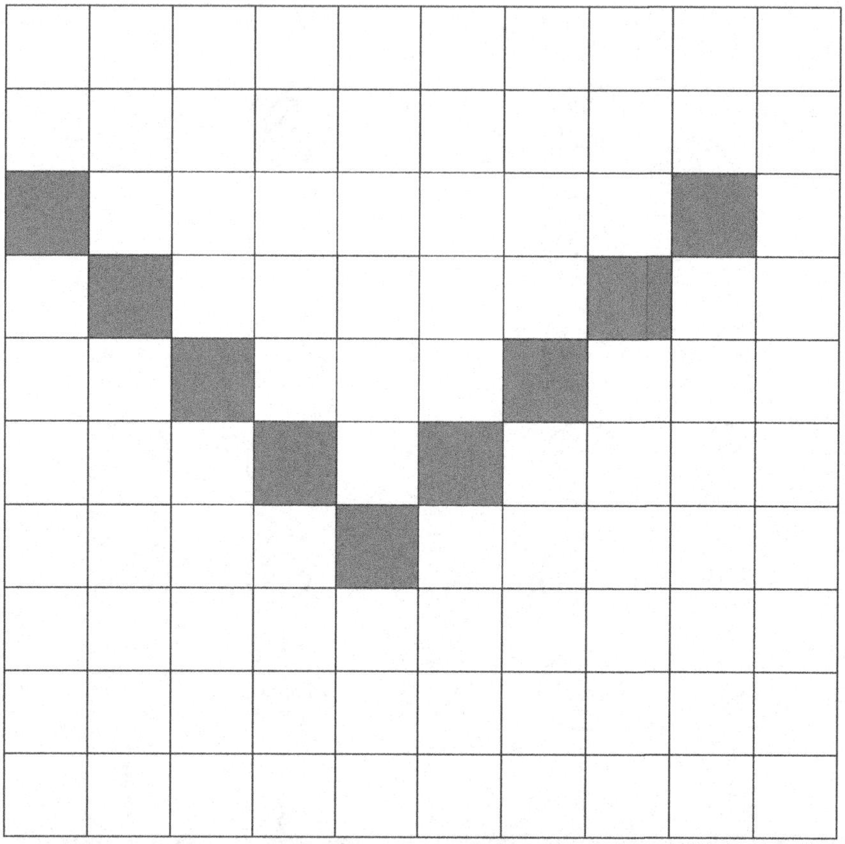

(A) 9

(B) 14

(C) 11

(D) 15

49. Daniel sorts his books as a group of four. He has twenty books in all. Which of the below shows his grouping as a multiplication fact ?

(A)

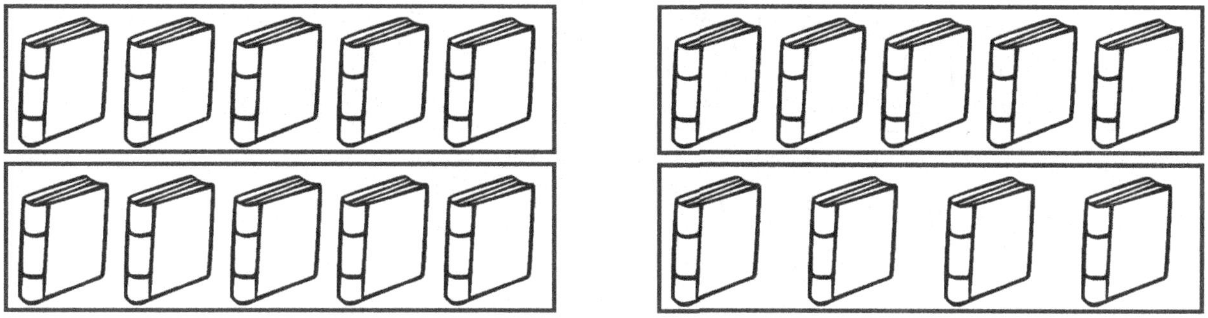

(B)

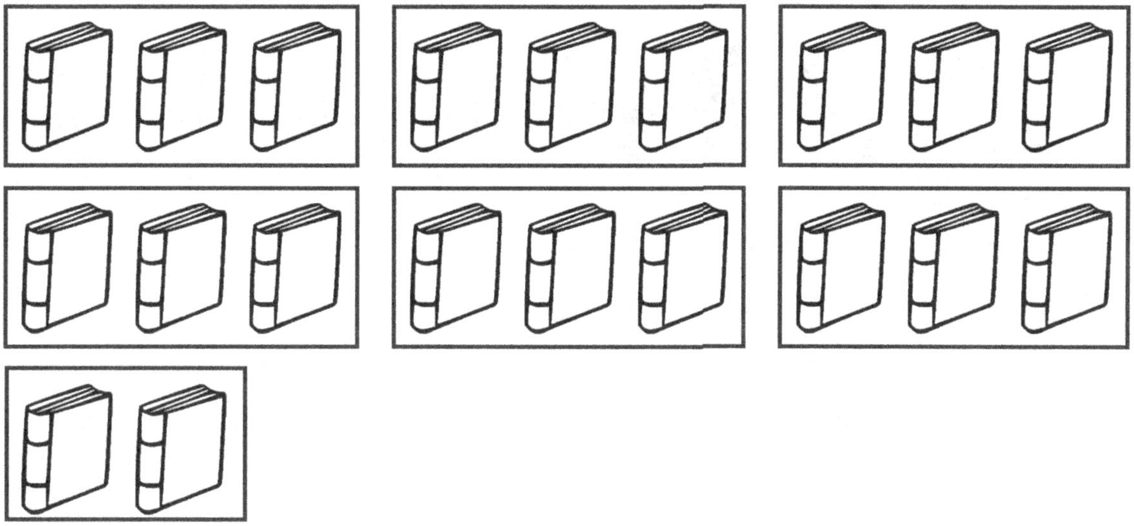

(C)

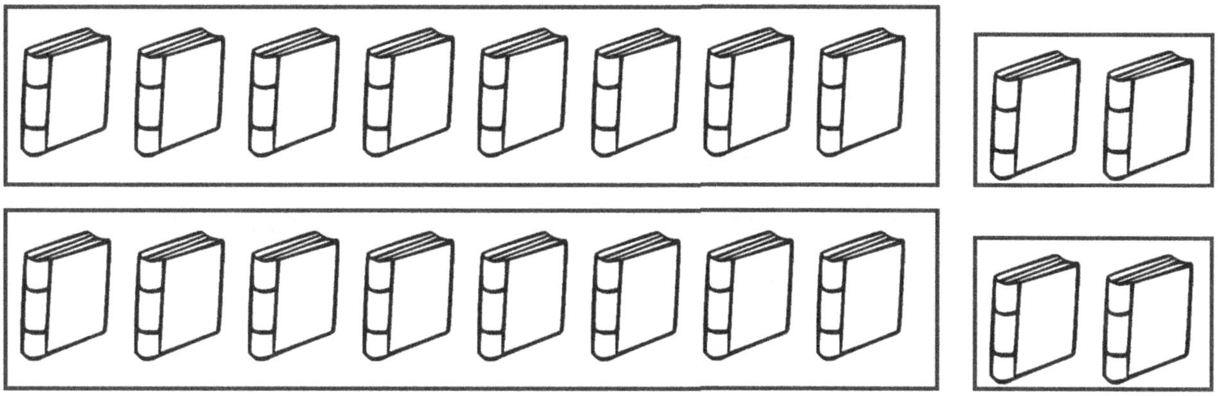

(D) 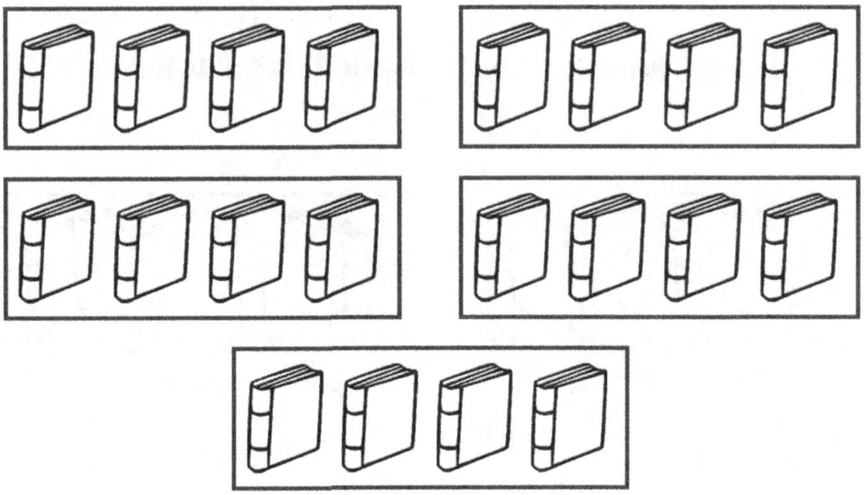

50. 

If the above pattern continues, what are the next two figures in the pattern ?

(A) 

(B) 

(C) 

(D) 

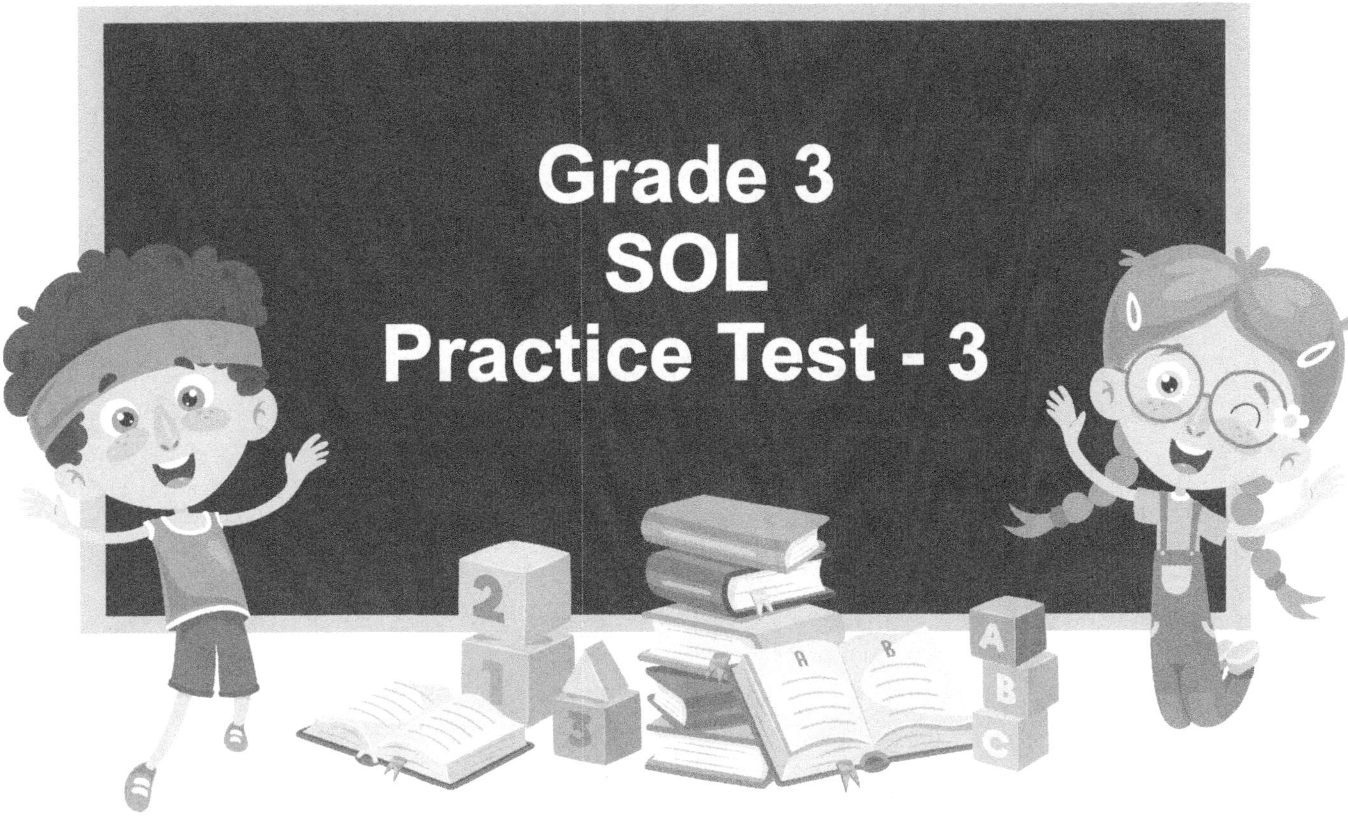
Grade 3
SOL
Practice Test - 3

1. Fan Club of Virginia sells 3,909 tickets for their fashion show. What is 3,909 rounded to the nearest ten ?

   (A) 3,900

   (B) 3,910

   (C) 3,909

   (D) 4,000

2. Which of the below shows "Two hundred fifteen thousand, nine hundred one" in standard form ?

   (A) 2,015,901

   (B) 205,901

   (C) 206,405

   (D) 215,901

3. Mary places sixteen blocks in groups of two. Which of the below shows the fact ?

   (A) 

   (B) 

   (C) 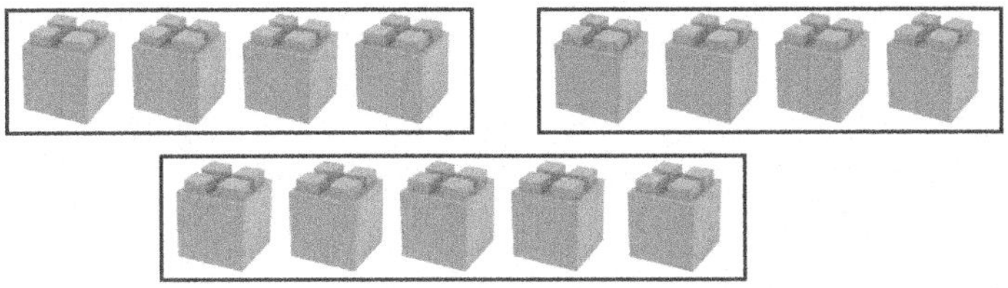

   (D) 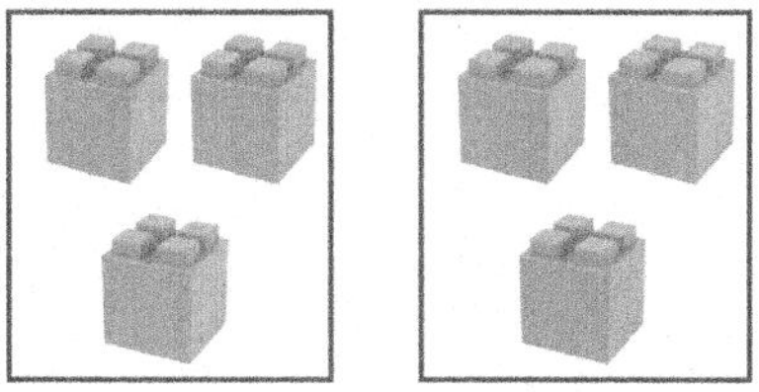

4. Which of the below statement is true ?

(A) $\frac{3}{5} < \frac{4}{9}$

(B) $\frac{3}{11} > \frac{3}{15}$

(C) $\frac{2}{7} > \frac{5}{7}$

(D) $\frac{2}{9} > \frac{5}{11}$

5. Which of the below statement is true ?

(A) 92,960 = 91,960

(B) 92,960 > 93,990

(C) 92,960 > 92,090

(D) 99,960 < 99,960

6. The below figure represents one whole part.

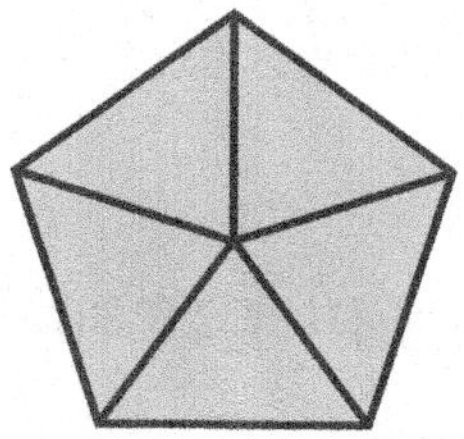

Then what number is represented by the below ?

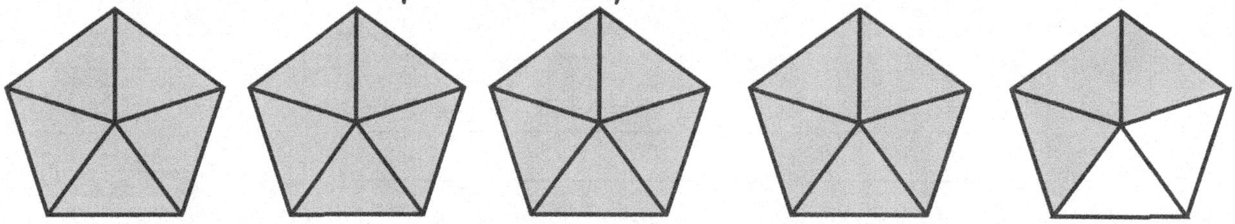

(A) $5\frac{2}{5}$ (B) $6\frac{3}{5}$ (C) $4\frac{3}{5}$ (D) $5\frac{3}{5}$

7. The figure below is shaded to represent the number 1.

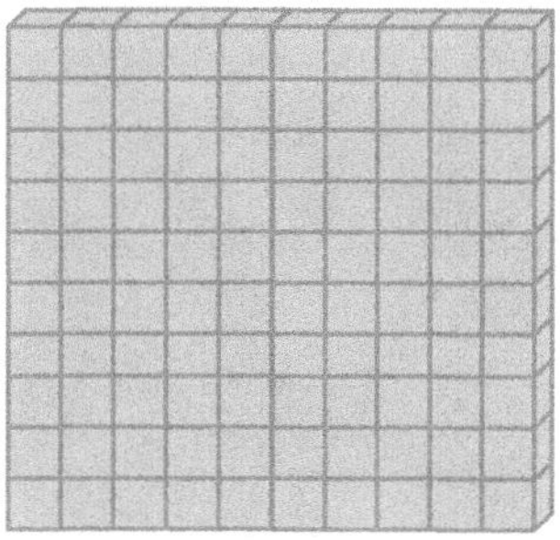

The below model represents which number ?

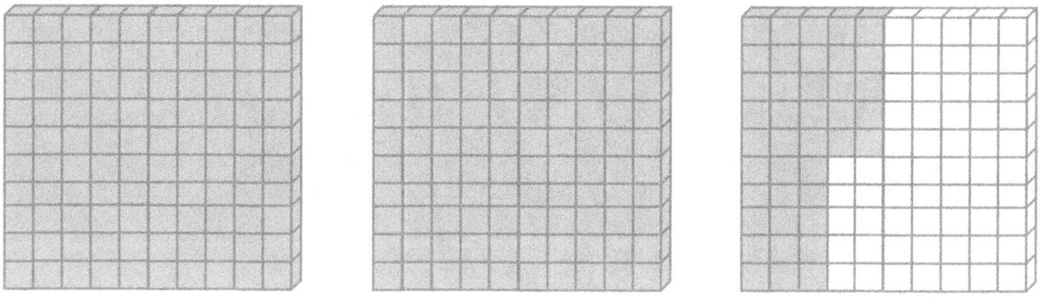

(A) 3.04 (B) 2.55 (C) 2.04 (D) 2.40

8. This model is shaded to show a fraction of a whole.

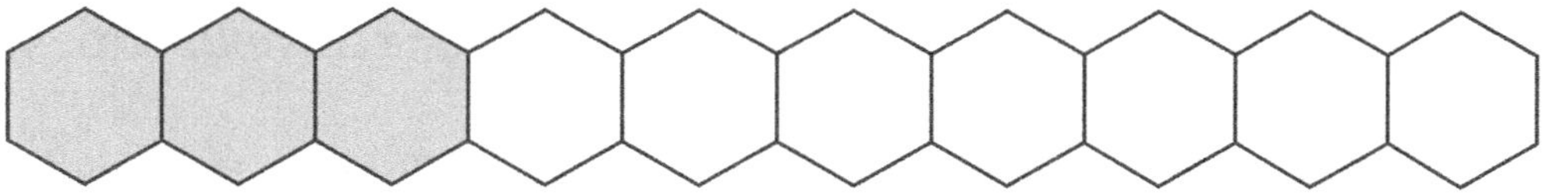

Which of the following is shaded to show a fraction with a value LESS THAN the one above ?

(A) 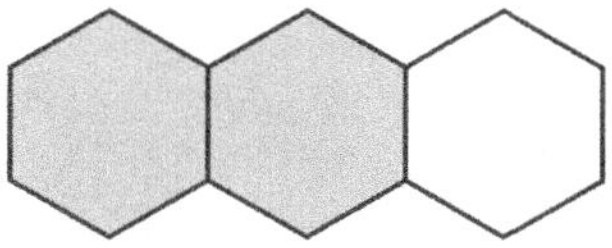

(B) 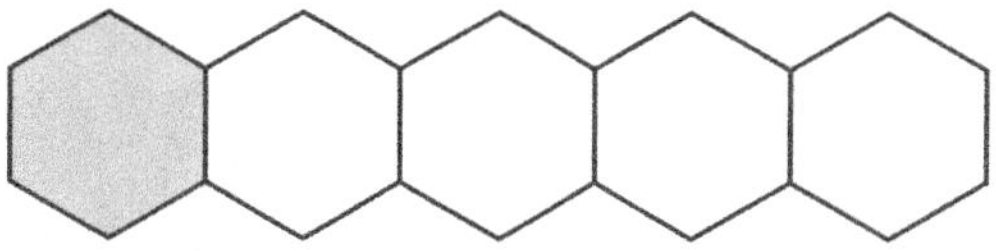

(C) 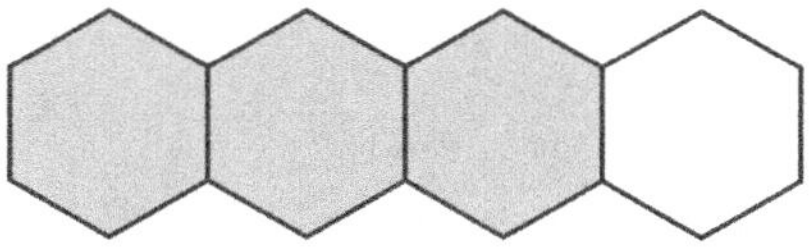

(D) 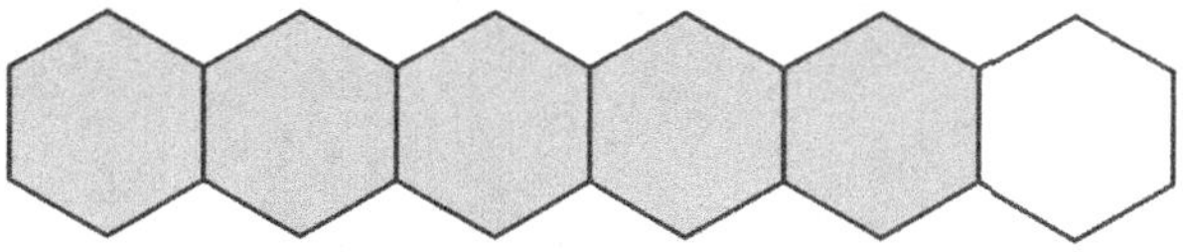

9. Which of the below box has more number of Stars ?

(A)

(B)

(C)

(D)

10. What is 5,093,075 rounded to the nearest hundred thousand ?

(A) 5,100,000

(B) 5,093,000

(C) 5,100,075

(D) 5,093,075

11. Ansh solved 98,051 math problems this year.
What is the value of 9 in 98,051 ?

(A) 900

(B) 9,000

(C) 90,000

(D) 9

12. This  represents a value of 1.

Which model represents the value of 1 in 777,109 ?

(A) 

(B) 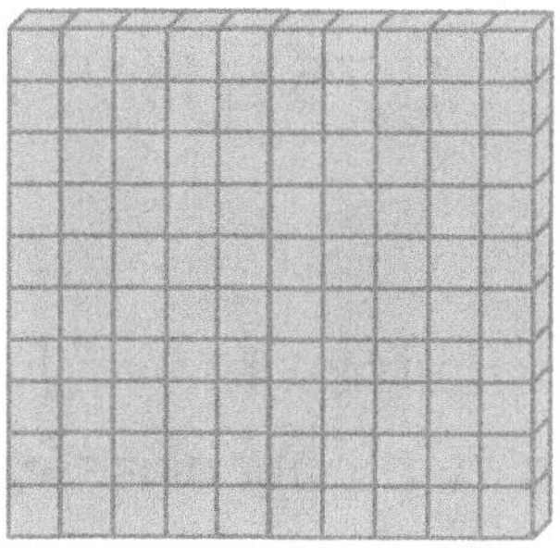

(C) 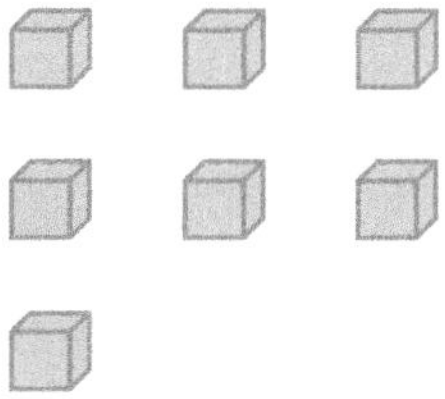

(D) 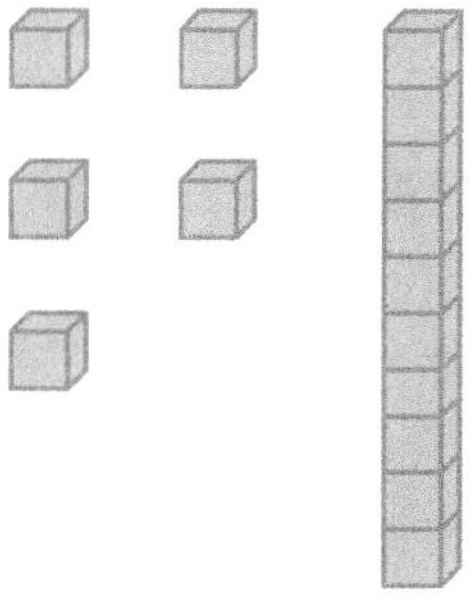

13. What fraction of triangles are there in total ?

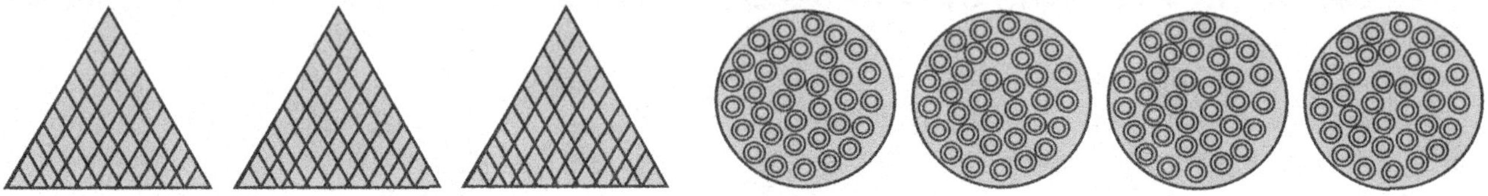

(A) $\frac{3}{7}$ (B) $\frac{3}{4}$ (C) $\frac{3}{5}$ (D) $\frac{3}{1}$

14. Which of the below statements is true ?

(A) 874 is less than 869

(B) 912 is less than 670

(C) 838 is equal than 1838

(D) 1275 is less than 999

15. Sri invites ten of her friends for her birthday party. She and her friends ate two cupcakes each. How many cupcakes did they ate in all ?

(A) 30

(B) 20

(C) 12

(D) 22

16. Evaluate $9 \times 5 = \boxed{?}$

(A) 36

(B) 18

(C) 45

(D) 95

17. Lisa has 24 hair clips and 15 hair bands

24	15
Hair clips	Hair bands

How many more hair clips does Lisa have than hair bands ?

(A) 39

(B) 9

(C) 12

(D) 17

18. Evaluate $1001 + 100 = \boxed{?}$

(A) 1,199

(B) 1,191

(C) 1,101

(D) 1,100

19. This model is shaded to show 1 whole.

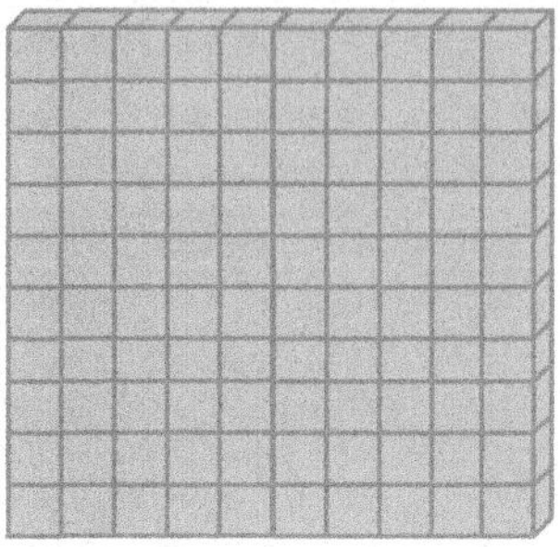

What decimal is shown by the model below ?

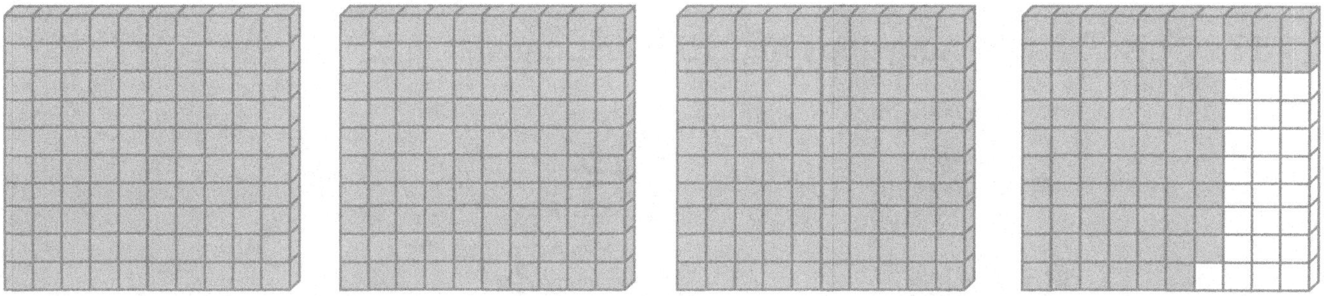

(A) 3.75 (B) 3.69 (C) 3.74 (D) 3.71

20. Evaluate

$$\begin{array}{r} 8\,.\,08 \\ +\ 1\,.\,23 \\ \hline \\ \hline \end{array}$$

(A) 12.3

(B) 10.9

(C) 9.31

(D) 25.8

21. This model is shaded to show one whole.

The two models below are each shaded to show a fraction

Model 1

Model 2

Model 3

All fractions are added together. Which model is shaded to show the total ?

(A) 1.8 (B) 2.2

(C) 1.9 (D) 2.0

22. Evaluate

10 x 0 = ?

(A) 10

(B) 0

(C) 01

(D) 90

23. The below number sentence shows a multiplication property.

$$10 \times 7 = 7 \times 10$$

Which of the following number sentence shows the same property as above ?

(A) $100 \times 0 = 10 \times 0$

(B) $12 = 6 \times 2$

(C) $8 \times 8 = 16 \times 4$

(D) $12 \times 15 = 15 \times 12$

24. Evaluate

$$4.4 + 3.3 = \boxed{?}$$

(A) 7.7 (B) 7.8 (C) 7.9 (D) 7.07

25. Which of the figures below appears to have a line of symmetry ?

(A)
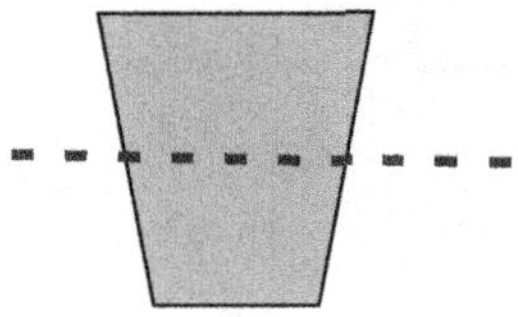

(B)
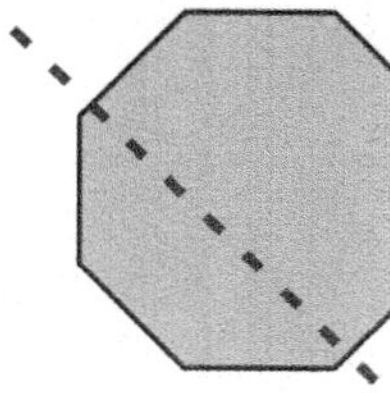

(C)
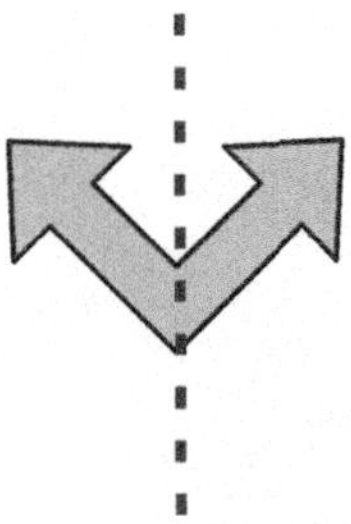

(D)
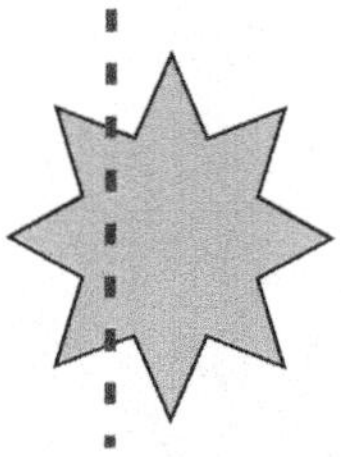

26. Alice bakes an apple pie and cuts it into 15 equal pieces. She gave 3 pieces to her sister Amy ,3 to her brother Andy and ate one piece herself. She saved rest of the pie for next day snack. What fractional part of the pie was left ?

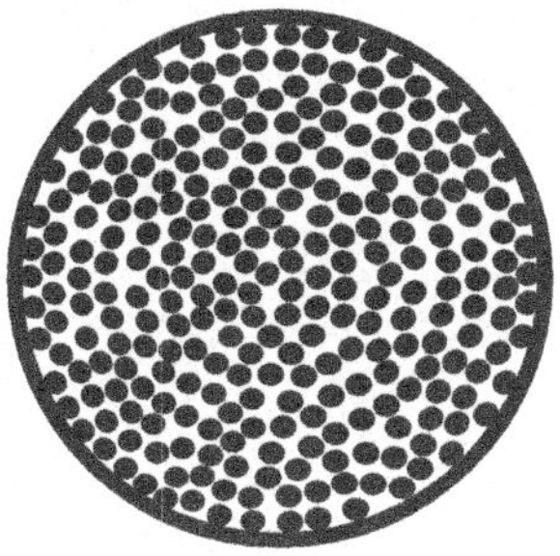

(A) $\frac{3}{15}$ (B) $\frac{8}{15}$ (C) $\frac{5}{15}$ (D) $\frac{9}{15}$

27. Weight of the truck is measured in which units ?

(A) Cups

(B) Gallons

(C) Miles

(D) Pounds

28. Jia worked for 2 hours today. How many minutes did she work today ?

(A) 120

(B) 150

(C) 60

(D) 90

29. Identify the closest temperature shown in the below figure ?

(A) 66^{0} Fahrenheit

(B) 150^{0} Fahrenheit

(C) 30^{0} Fahrenheit

(D) 13^{0} Fahrenheit

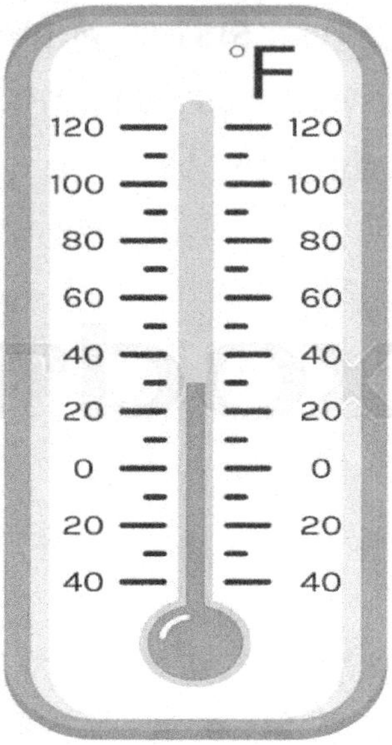

30. What number needs to go in the empty box to make the below number sentence true ?

$$37 + \square = 61$$

(A) 98 (B) 24 (C) 4 (D) 51

31. What is the closest length of the backpack ?
(Ruler is given in the right side for measurement)

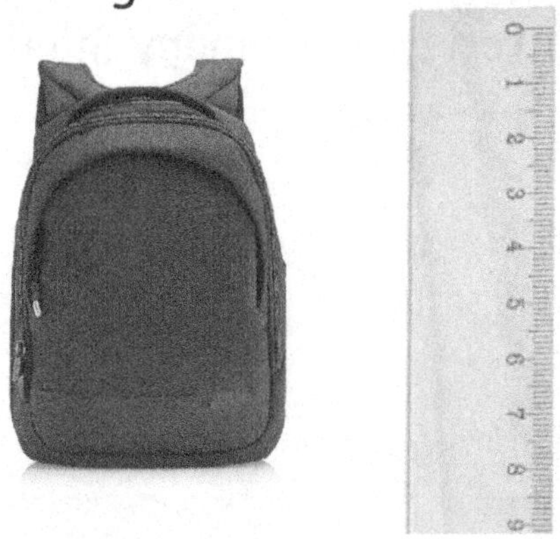

(A) 12 cm (B) 10 cm (C) 5 cm (D) 8 cm

32. Ben sorts his toy cars as a pair. He has fourteen cars in all. Which of the below shows his grouping as a multiplication fact ?

(A) 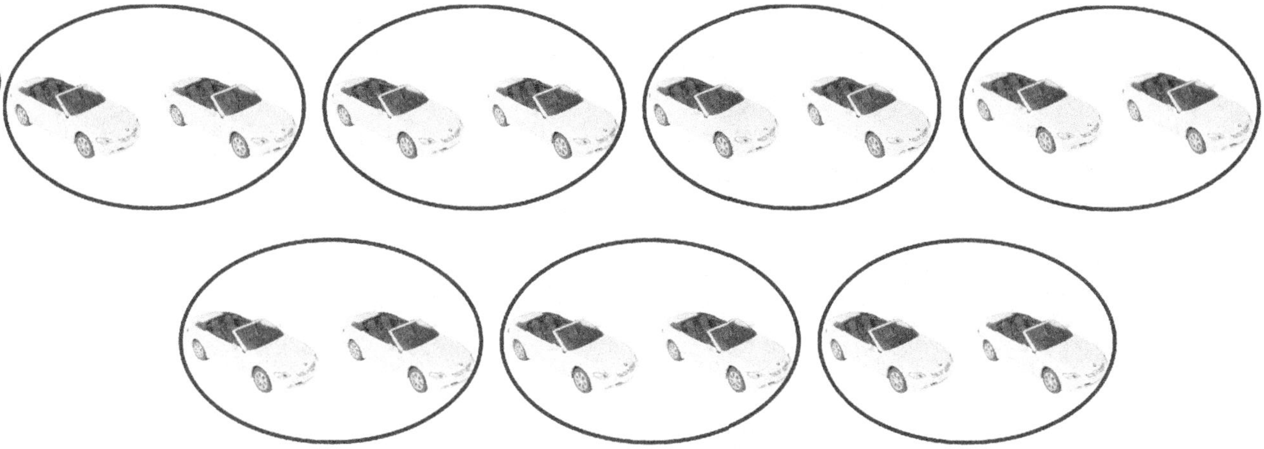

(B) 

(C) 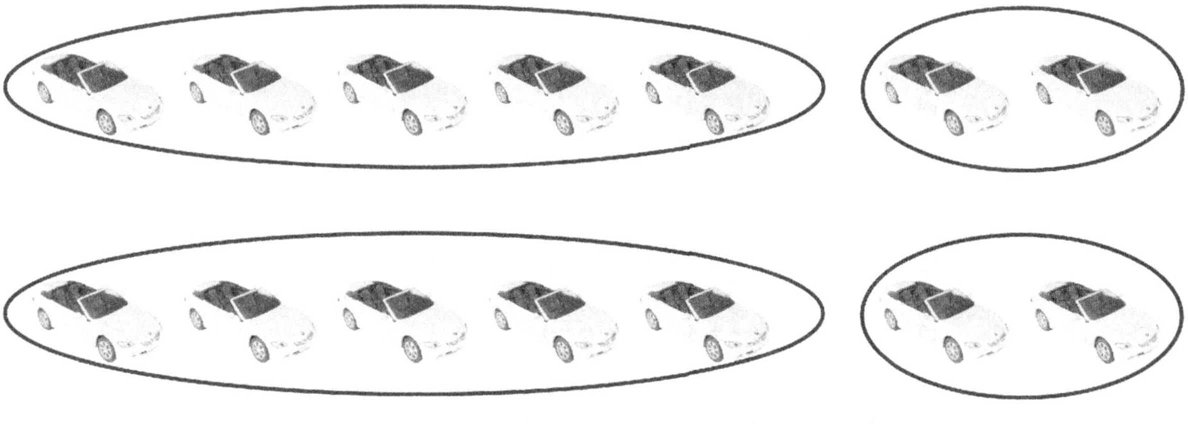

(D) 

33. Identify which of the following figures has the line of symmetry across the dotted line ?

(A) 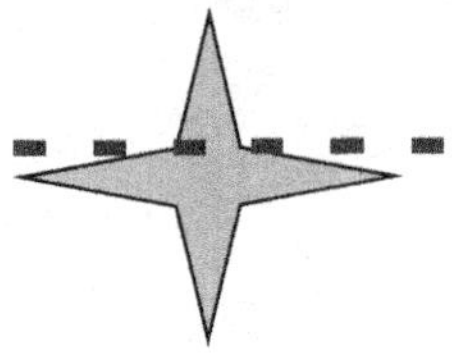

(B) 

(C) 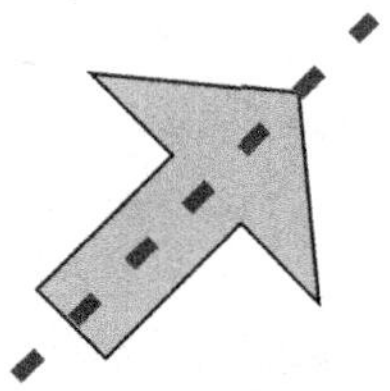

(D) 

34. A Parallelogram has ......?

(A) All sides are equal

(B) Two pairs of parallel sides

(C) Opposite sides are not parallel

(D) All angles are 90 degrees

35. Henry has

He saved this week

How much money did he have in all ?

(A) $ 9.30

(B) $ 8.30

(C) $ 8.99

(D) $ 8.03

36. What is the perimeter of an equilateral triangle as given below ?

(Approximately using the ruler)

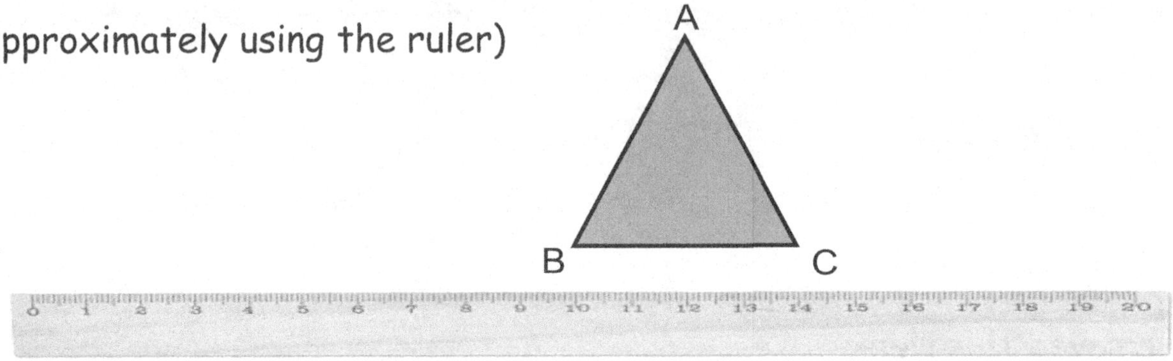

(A) 12 inches (B) 17 inches (C) 10 inches (D) 16 inches

37. Below data shows the number of puzzles solved by G & M math club this quarter ?

Month	January - 2019	February - 2019	March - 2019
Solved Puzzles	99	110	77

He prepares the below chart representing the above data

Month	Number of puzzles solved
January - 2019	★★★★★★★★★
February - 2019	★★★★★★★★★★
March - 2019	★★★★★★★

The symbol ★ represents the value of ___________ ?

(A) 12 (B) 15 (C) 11 (D) 10

38. Anna organizes a sports equipment collection drive for her school. She receives various sports balls as below.

Which bar graph shows the number of various balls received as part of the sports equipment collection drive ?

(A)

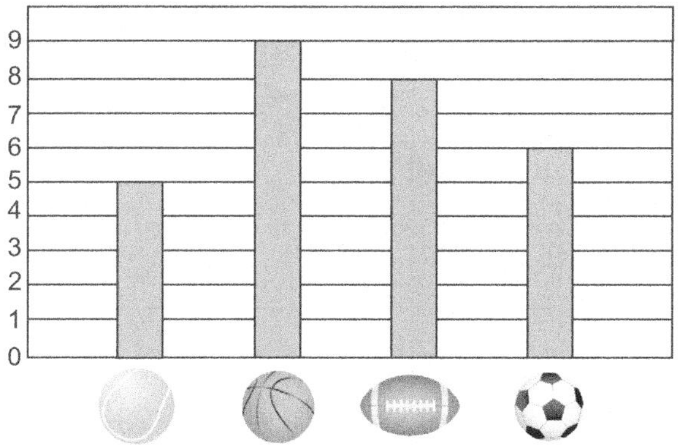

(B)

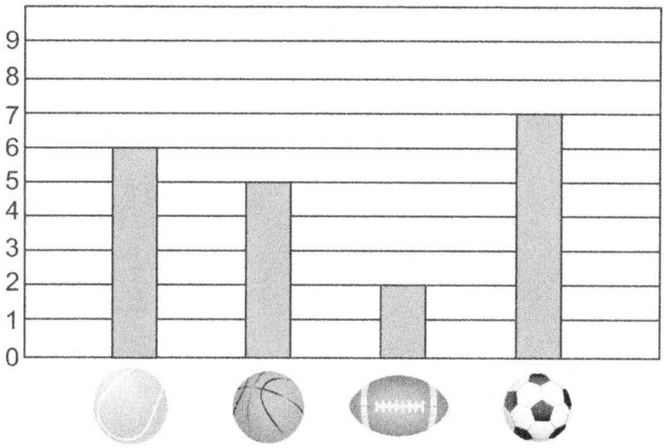

(C)

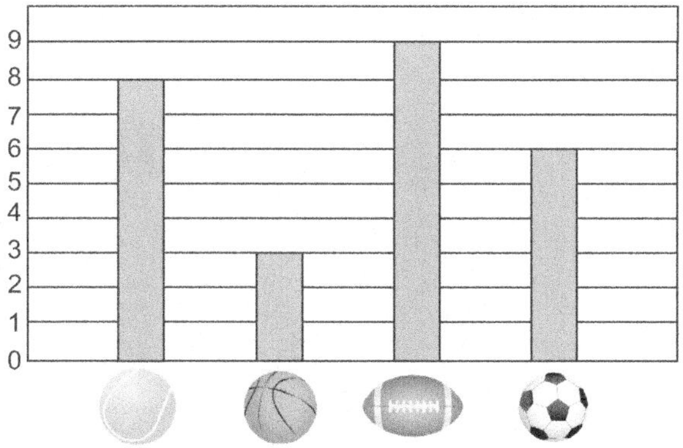

(D)

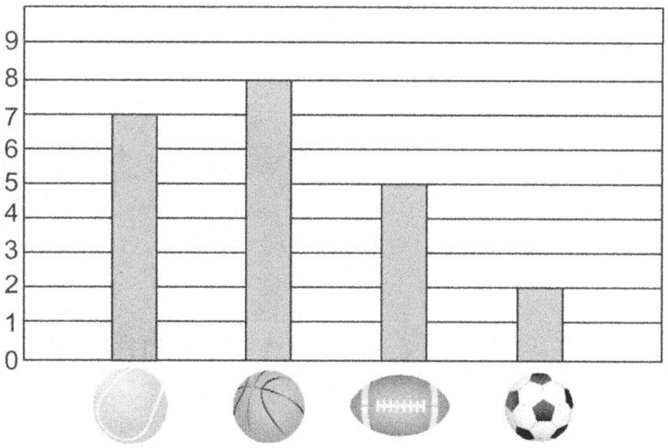

39. Maria picks a number from the number box below. What is the probability of picking a two digit number ?

0 3 13 7 20 9
8 18 16 5
15 2 14
4 19 17 1
11 6 12

(A) Certain

(B) Likely, but not certain

(C) Unlikely, but not impossible

(D) Impossible

40. The wakey weather company predicts on an average record low temperatures over the next few weeks. Below graph shows weeks horizontally. What is missing vertically ?

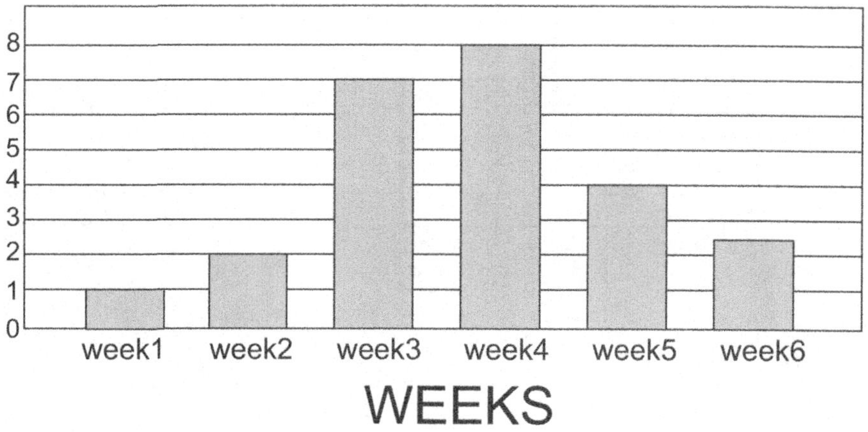

(A) Average temperature

(B) Lowest temperature

(C) Total temperature in a week

(D) Median temperature

41. Alice colored small squares in the shape of an alphabet from the below grid. If ☐ is equal to one square centimeters. What is the area of the alphabet "G" ?

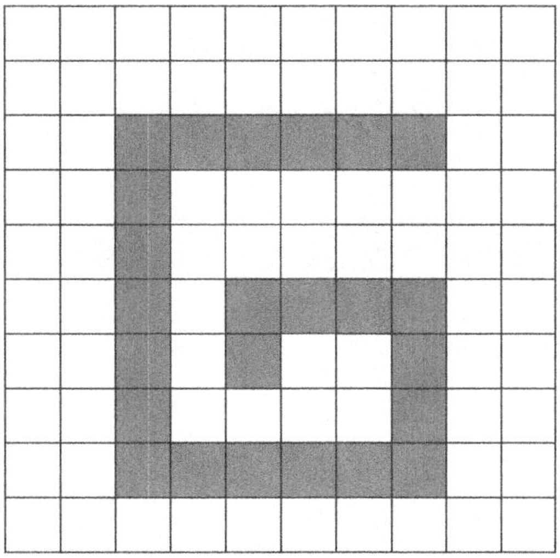

(A) 12 (B) 18 (C) 32 (D) 24

42. Find the next two numbers in the series ?

7 , 14 , 21 , 28 , 35,.....?

(A) 35 , 42 (B) 42 , 49 (C) 49 , 54 (D) 50 , 49

43. Which number can be placed on the line to make the number sentence true ?

10 X ? = 100

(A) 5 (B) 10 (C) 15 (D) 20

44. Using the skip counting rule what is the next number in the series ?

1000 , 1005 , 1010 , 1015 , 1020 , 1025,... ?

(A) 1035 (B) 1020 (C) 1030 (D) 1540

45. Jolly and her friends are doing a bake sale for a non profit organization. The table below shows the cost of cup cakes they are selling.

Price of Cup cakes

Number of Cup Cakes	Total Cost
1	\$ 2.10
2	\$ 4.20
3	\$ 6.30
5	\$ 10.50
10	\$?

Based on the pattern in the table, what will be the total cost for ten Cup Cakes ?

(A) \$ 21.10

(B) \$ 15.70

(C) \$ 12.60

(D) \$ 21.00

46. Mona has water bottles, lunch box and spoons in her trunk. She counts them after sorting. Which chart below shows the correct count ?

(A)

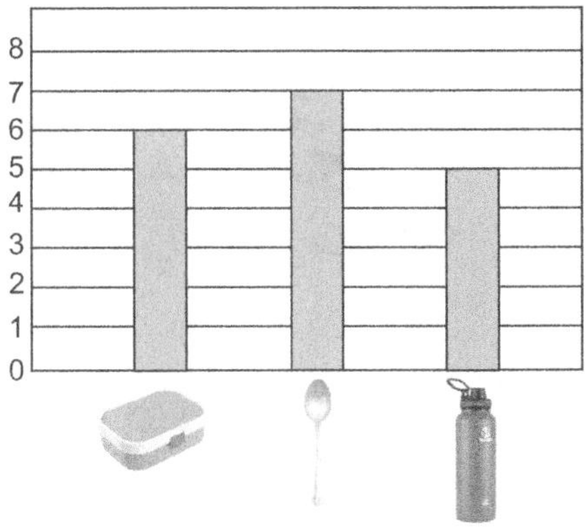

(B)

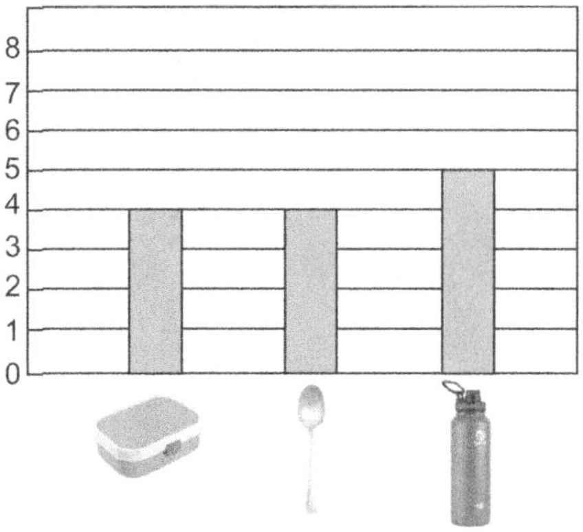

(C)

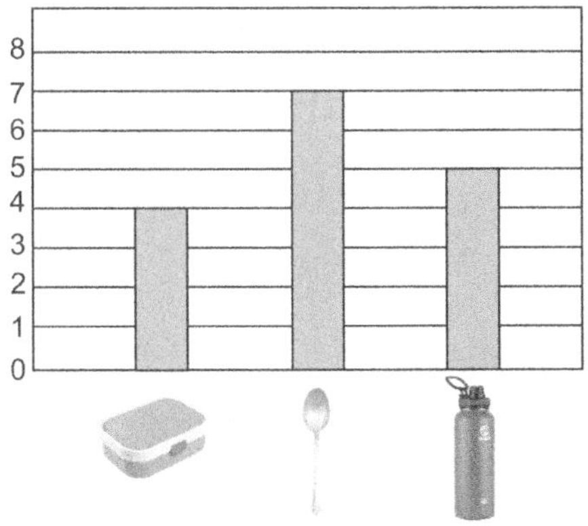

(D)

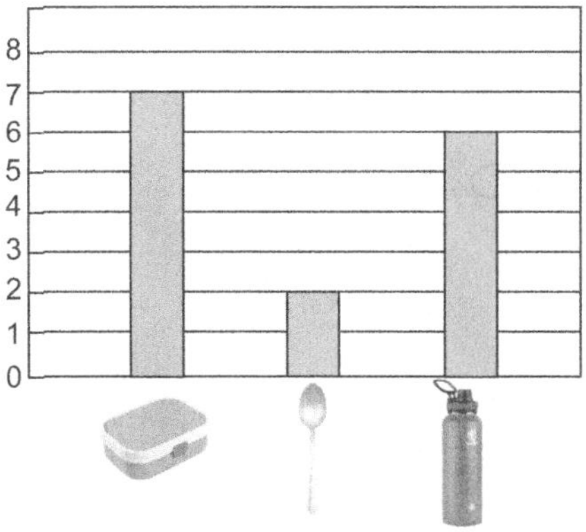

47. Which part of the below picture represents a triangle ?

(A) Roof

(B) Door

(C) Tree

(D) Windows

48. Kate is wondering which of the below is showing the time of 9.45PM ?

(A) 

(B) 

(C) 

(D) 

49. Beth uses the math fact $\boxed{13 \times 4 = 52}$ to solve a math problem. Which of the below could be the problem she is trying to solve.

(A) $13 - \square = 8$

(B) $\square \times 13 = 65$

(C) $\square + \square = 13$

(D) $\square \div 4 = 13$

50. The number of students voted for their favourite disney character. The data is plotted as below.

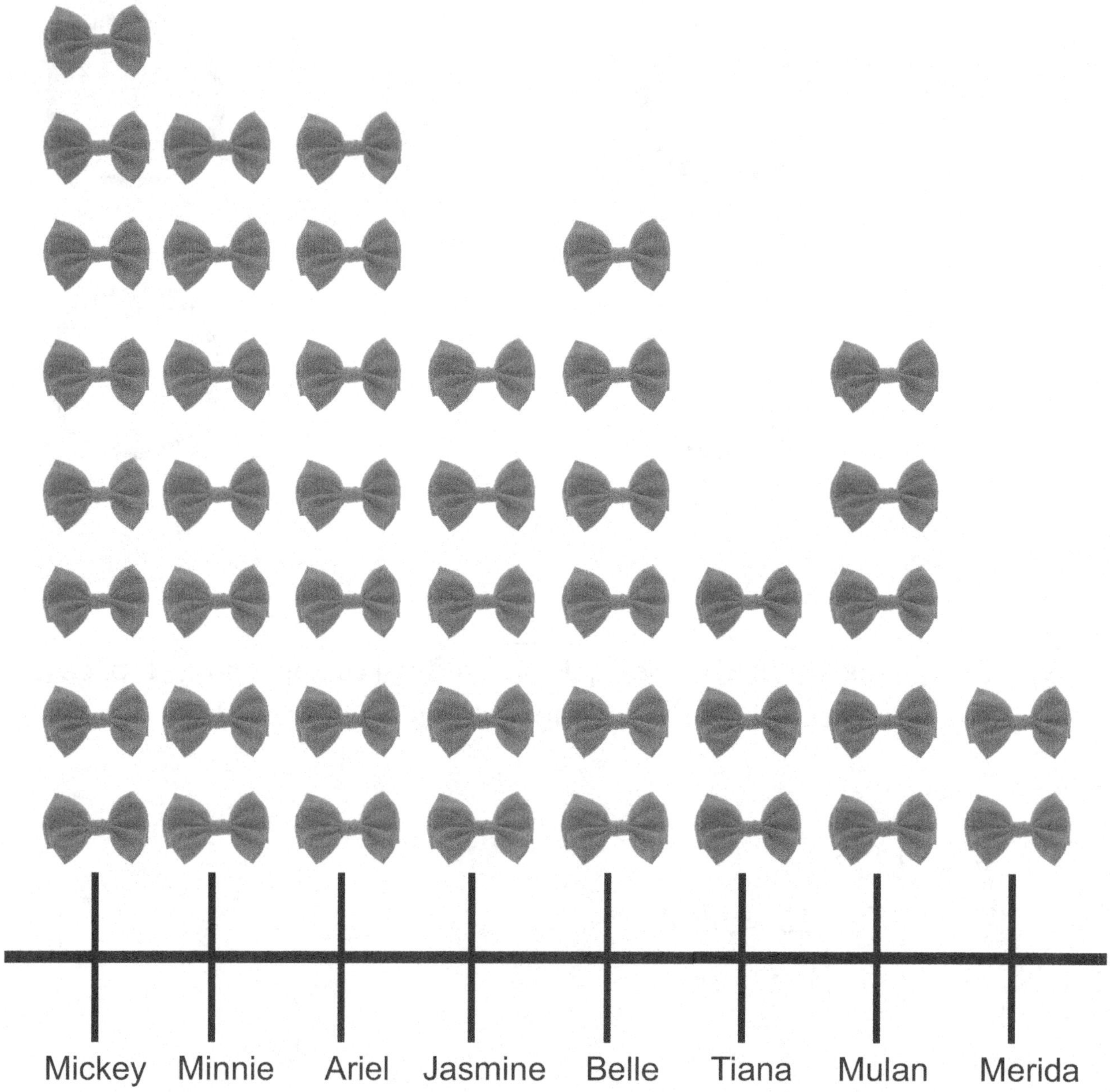

Which two characters have same number of votes ?

(A) Minnie & Belle

(B) Minnie & Ariel

(C) Jasmine & Tiana

(D) Mickey & Minnie

GRADE 3
SOL
Practice Test - 1
Answer Keys

Answer Key Test - 1

1. D
2. A
3. C
4. B
5. D
6. B
7. C
8. D
9. A
10. C
11. D
12. B
13. A
14. D
15. C

16. B

17. B

18. D

19. C

20. A

21. D

22. B

23. C

24. A

25. D

26. C

27. B

28. C

29. A

30. C

31. B

32. C

33. C

34. D

35. B

36. D

37. A

38. C

39. D

40. D

41. A

42. A

43. C

44. B

45. C

46. B

47. D

48. B

49. D

50. A

GRADE 3
SOL
Practice Test - 2
Answer Keys

Answer Key Test - 2

1. A
2. D
3. B
4. C
5. A
6. C
7. A
8. D
9. C
10. C
11. A
12. B
13. A
14. C
15. C

16. B

17. C

18. A

19. A

20. D

21. A

22. C

23. B

24. D

25. A

26. C

27. B

28. D

29. B

30. A

31. B

32. D

33. A

34. A

35. B

36. A

37. D

38. B

39. C

40. A

41. B

42. D

43. C

44. D

45. A

46. B

47. D

48. A

49. D

50. C

GRADE 3
SOL
Practice Test - 3
Answer Keys

Answer Key Test - 3

1. B
2. D
3. A
4. B
5. C
6. C
7. D
8. B
9. D
10. A
11. C
12. B
13. A
14. A
15. D

16. C

17. B

18. C

19. A

20. C

21. D

22. B

23. D

24. A

25. C

26. B

27. D

28. A

29. C

30. B

31. D

32. A

33. C

34. B

35. D

36. A

37. C

38. A

39. B

40. A

41. D

42. B

43. B

44. C

45. D

46. C

47. A

48. C

49. D

50. B

GRADE 3
SOL
Score Calculation

Score calculation

If you get this many times correct :	Then your converted scale scor is :
0	000
1	217
2	249
3	268
4	282
5	293
6	303
7	311
8	319
9	325
10	331
11	337
12	343
13	348
14	353
15	357
16	362
17	366
18	371
19	375

20	379
21	383
22	387
23	391
24	395
25	399
26	403
27	407
28	411
29	415
30	419
31	423
32	427
33	432
34	436
35	441
36	445
37	450
38	456
39	461

40	467
41	473
42	480
43	487
44	496
45	505
46	516
47	531
48	550
49	582
50	600

Made in United States
North Haven, CT
06 November 2026

21791726R00065